Handbook of

Fire Protection Engineering

Handbook of
Fire Protection Engineering

K. Ehteshami, *P.E.*
Professional Engineer,
University of Texas (U.S.A)

KHANNA BOOK PUBLISHING CO. (P) LTD.

Publisher of Engineering and Computer Books

4C/4344, Ansari Road, Darya Ganj, New Delhi-110002
Phone : 011-23244447-48 **Mobile:** +91-9910909320
E-mail : contact@khannabooks.com
Website : www.khannabooks.com

Dear Readers,

To prevent the piracy, this book is secured with HIGH SECURITY HOLOGRAM on the front title cover. In case you don't find the hologram on the front cover title, please write us to at contact@khannabooks.com or whatsapp us at +91-99109 09320 and avail special gift voucher for yourself.

Specimen of Hologram on front Cover title:

Moreover, there is a SPECIAL DISCOUNT COUPON for you with EVERY HOLOGRAM.

How to avail this SPECIAL DISCOUNT:

Step 1: Scratch the hologram
Step 2: Under the scratch area, your "coupon code" is available
Step 3: Logon to www.khannabooks.com
Step 4: Use your "coupon code" in the shopping cart and get your copy at a special discount
Step 5: Enjoy your reading!

Price: ₹ 495.00
$ 18.00

ISBN: 978-93-86173-77-5

Handbook of Fire Protection Engineering *by* K. Ehteshami

First Edition: 2019
Reprint Edition: 2023, **2025**

To view complete list of books, Please scan the QR Code:

Published by:
Khanna Book Publishing Co. (P) Ltd.
CIN: U22110DL1998PTC095547

Visit us at: **www.khannabooks.com**
Write us at: **contact@khannabooks.com**

Printed in India.

To my father

Masoud Khan Ehteshami

and

my mother

Bi Bi Shamsolmolook Ilbag

and

to my nanny

Mashhadi Jamileh Farhadi

May all Rest in Peace

PREFACE

The science of Fire Protection Engineering has gone through a great deal of changes for the past 100 years. Technological advances introduced new hazards in the workplace, caused new regulations and laws governing the design and installation of fire protection systems. The fire protection technology along with new regulations and laws have made the job of fire protection professionals more challenging and important. This also created the need for more up-to-date book containing the latest information in technology and methods to aid and enable the fire re protection professionals to compete in the ever changing and challenging world.

I wrote this book in response to the need for an up-to-date practical handbook which I believe is highly useful in providing help for engineers, architects, contractors, colleges, universities and regulatory agencies. The direct and complete presentation of materials in this handbook are specially useful in the design and installation of modern fire protection systems. Also there is a chapter on explosion prevention which is highly useful in woodworking shops, munition making factories, etc... .

Each chapter in this book focuses on the major area of concern of fire protection professionals. The chapters are organized and presented in the order that is compatible to engineering text books.

ACKNOWLEGEMENT

NFPA (National Fire Protection Association) pamphlets and my own 42 years of experience as Project Engineer and Project Manager supervising design, fabrication and installation of fire protection systems for all types of hazards in all kind of buildings such as mid-rise and high-rise office and apartment buildings shopping centers, hospitals and nursing homes. manufacturing plants and warehouses, airports, etc...

CONTENTS

1 Introduction to Fire Protection

The conditions necessary for the existence of a fire are the presence of fuel (a combustible substance), heat (a temperature high enough to cause combustion), and oxygen (usually provided by air).

Fire fighting consists of removing one or more of the three elements essential to combustion: fuel, heat, or oxygen. Different types of fuel ignite at different temperatures. The ignition temperature of a combustible solid is influenced by the rate of the airflow, the rate of heating, and the size of the solid.

Most firefighting consists of applying water to the burning materials, thereby cooling it down to the point at which combustion is no longer self-sustaining. Fires involving flammable liquids, certain chemicals, and combustible metals often require special extinguishing agents and techniques. With some fuels the use of water may actually be dangerous.

In the preindustrial era, most cities had watchmen who looked for signs of fires and sounded an alarm when necessary. The Roman emperor Augustus is credited with instituting a corps of fire-fighting watchmen. In ancient Rome, as in the Middle Ages and early modern times, the principal piece of fire-fighting equipment was the bucket, passed from hand to hand to carry water from a well or river to the fire area. Another piece of primitive fire-fighting equipment was the hatchet or axe, used to remove the fuel and prevent the spread of fire as well as to allow heat and smoke to escape.

The basic tactics of fighting a fire can be divided into the following categories: rescue operations, protection of buildings exposed to the fire, confinement of the fire, extinguishing the fire, and salvage operation.

Some of the best fire-fighting equipment and methods are: fire alarms, smoke detectors, automatic fire sprinkler systems, nozzles, water additives, and fire prevention.

Alarms are transmitted to a fire station by radio or over special alarm circuits. In large cities, alarm are received at a central dispatch office and then transmitted to fire stations, frequently with the use of mobile teleprinters and computers.

Buildings may also be equipped with smoke detectors, and in many cities such systems are now required by law in public buildings and apartment houses. Two major types of smoke detectors are available. One is an ionization device that contains a small radioactive source for ionizing the air molecules between a pair of electrodes, permitting a very small current to flow between the pair. If smoke particles from a fire enter this space, they reduce the flow of current by adhering to the ionized molecules. The drop in current sets off a buzzer or other alarm. The second type of smoke detector uses a photoelectric cell. In some of these detectors, smoke that enters obscures a steady beam of light; in others, the smoke scatters a light ray from a diode so that the cell can detect it. In either case the change sets off an alarm. Both types can be run by batteries or by a building current.

Automatic sprinkler systems detect fire by means of heat-sensitive or other devices, which then trigger the sprinkler to put out the flames. Sprinkler systems are nearly 100 percent effective. Many are supervised electrically from a central station, and alarms are transmitted to a fire department whenever the sprinklers operate or when a valve in the sprinkler system closes for any reason. An automatic sprinkler system is one of the most effective means of fire protection and firefighting today. An electrically supervised sprinkler system is even more effective, since it provides a far greater range of protection against sprinkler failure and tampering. In addition, its modular design allows easy expansion to provide detection and notification by zones as well as to interface with other building control systems.

Various nozzles are capable of projecting a solid, heavy stream of water, curtains of spray, or fog. Fire trucks carry a selection of nozzles, which are used according to the amount of heat that must be absorbed. Nozzles can apply water in the form of streams, spray, or fog at rates of flow between 15 gallons and more than 100 gallons per minute. Straight streams of water have a greater reach and penetration, but fog absorbs heat more quickly because the water droplets present a greater surface area and distribute water more widely. Fog nozzles are usually used to control fires involving flammable liquids, although foam equipment is also used for such a fire.

Fire fighters often add various chemicals to water to improve its ability to extinguish fires. Wetting agents are sometimes added to the water to reduce the surface tension of the water. This makes the water more penetrating and facilitates the formation of small drops necessary for rapid heat absorption. By adding foam-producing chemicals and liquids to water, a fire-blanketing foam is produced. Foam is used to extinguish fires in combustible liquids, such as oil, petroleum, and tar, and for fighting fires at airports, refineries, and petroleum distribution facilities.

Perhaps ever more important than firefighting itself is fire prevention. Fire prevention bureau in the fire department of most major cities directs fire prevention activities. It handles the more technical fire prevention problems, maintains appropriate records, investigates the causes of fires, and conducts public education programs. It also inspects buildings and reviews plans for code compliance.

2 Antifreeze System

Introduction

An antifreeze sprinkler system is a fixed fire protection system using piping filled with an antifreeze solution and connected to a pressurized water system. The system is activated by fusible sprinkler heads which go off at a pre-set temperature. The antifreeze, followed by water, discharges from sprinkler heads and sprays the fire area. This action would put the fire out or hold it in check until firemen arrive.

An antifreeze sprinkler system can be installed in any building with a small unheated area. The building must be structurally sound to support the piping system filled with water. The low cost of equipment, installation, and maintenance plus the speed of operation would make an antifreeze system preferable to a dry pipe system.

Components of the System

1. Water Supply
   An adequate water supply taken from the city main, gravity tank, pressure tank, storage tank, and well.
2. Underground System
   a. Piping – Cast iron, ductile iron, cement, or plastic.
   b. Control valves - Post indicator valve (PIV).
   c. Valve pit – Two Outside Stem & Yoke (OS & Y) gate valves, one Check Valve (CV) or detector check valve, a Fire Department Connection (FDC), a check valve with ball drip, a backflow preventer, a full flow meter, or a combination of equipment may be locally required.
   d. Fire hydrants – Consult with the local fire department for location and thread requirements.
3. Overhead System
   a. A water control valve.
   b. Two solution test valves.
   c. A check valve with a 1/32 inch hole in the clapper.

d. A filling cup.
e. Non-freezing solution.
f. A system of piping which carries the antifreeze solution and water to the sprinklers. The pipe size is determined by hydraulic calculations.
g. Sprinkler type, temperature rating, orifice size, and coverage is determined by the design criteria or recommended by the authorities having jurisdiction.
h. Hangers and earthquake bracing are required. The type and spacing is determined by the design criteria, or recommended by the authorities having jurisdiction.
i. A drain valve is required on each system.

System Operation

In case of a fire, the heat will fuse sprinklers and that causes the flow of antifreeze solution followed by water.

3 Deluge System

Introduction

A deluge system is a fixed fire protection system which totally floods the area of fire with pressurized water through a system piping leading to open nozzles or open sprinklers. The piping system is empty until the deluge valve is activated by a hydraulic, pneumatic, electric, or manual release.

Deluge systems may be installed in extra hazard occupancies to create a buffer zone or to cool a structure to prevent building collapse. They are used in storage or process areas having a low flash point, tanks containing combustible solutions, transformers, and loading docks for garment companies. A foam-water deluge system utilizes foam- water nozzles or sprinklers to smother and extinguish fires in aircraft hangars, extraction plants, and areas where flammable liquids are used.

Components of the System

1. The Design Problems
   a. Type of hazard.
      In order to apply the appropriate codes and standards, the specific information about the hazard of the area to be protected must be obtained.
   b. System's purpose.
      A system should be designed to cool surfaces, to extinguish a fire, or to do both.
   c. Area to be protected.
      The maximum flow for a deluge system is 3,000 gpm.
   d. Hater supply.
      Hater demand must be hydraulically calculated and an adequate water supply taken from the city main or an existing system.
   e. Drainage.
      A deluge system delivers a very large volume of water to the fire area. The building owner must be notified of the amount of water his drain system has to drain so he can take appropriate measures to insure that his drain system has the capacity to drain the water.

f. Combustible liquids.

If there is floatable combustible liquid in the protected area, an adequate drainage or diking system should be installed to insure that fire will not be transferred to adjacent areas.

2. Underground System

a. Piping – Cast iron, ductile iron, cement, or plastic.

b. Control valves – Post indicator valve (PIV).

c. Valve pit – Two Outside Stem & Yoke (OS & Y) gate valves, one Check Valve (CV) or detector check valve, a Fire Department Connection (FDC), a check valve with ball drip, a back- flow presenter, a full flow meter, or a combination of equipment may be locally required.

d. Fire hydrants – Consult with the local fire department for location and thread requirements.

3. Overhead System

a. Wall PIV, OS & Y gate valves, or other type of control valves, if it is not provided in the underground system.

b. A check valve, if it is not provided in the underground system.

c. Deluge valve with all the trimming and release equipment.

d. Emergency release.

e. System piping.

f. Release system piping.

g. Water motor alarm.

h. Spray nozzles.

i. Thermostatic release.

System Operation

When the pilot head or emergency release is activated, water escapes from the top chamber of the deluge valve faster than it can be replaced. This causes the deluge valve to open, and pressure builds up in the system side of the deluge valve and opens the Pressure Operated Release Valve (PORV). Water rushes through the piping system and discharges from open sprinklers or nozzles.

4 Dry Pipe System

Dry Pipe Sprinkler System

A dry pipe sprinkler system is a fixed fire protection system which utilizes pressurized water as an extinguishing agent. All the piping from the dry pipe valve to the sprinklers is filled with pressurized air or other gases such as nitrogen.

A dry pipe system is primarily used to protect unheated buildings, where temperature falls below the freezing point.

The building must be structurally sound to support the system piping filled with water. One dry pipe system can protect an area no larger than 52,000 square feet.

A dry pipe system has the same floor area and fire limitations as a wet pipe system. In a dry pipe system, the sprinklers are installed in an upright position. If pendent sprinklers are needed, they must be of a dry pendent type to prevent the freezing of the trapped water.

Dry pipe systems operate slower than wet pipe systems. Therefore, they must be installed where life safety is not a prime concern.

Components of the System

1. Water Supply
   An adequate water supply taken from the city main, gravity tank, pressure tank, storage tank, and well.
2. Underground System
   a. Piping – Cast iron, ductile iron, cement, or plastic.
   b. Control valves – Post Indicator Valve (PIV).
   c. Valve pit – Two Outside Stem & Yoke (OS & Y) gate valves, one Check Valve (CV) or detector check valve, a Fire Department Connection (FDC), a check valve with ball drip, a back flow preventer, a full flow meter, or a combination of equipment may be locally required.
   d. Fire hydrants – Consult with the local fire department for location and thread requirements.

3. Overhead System
   a. Wall PIV, OS & Y gate valves, or other type of control valves, if it is not provided in the underground system.
   b. A check valve, if it is not provided in the underground system.
   c. Dry pipe valve with the following trimming:
      i. The dry pipe valve and all the piping leading to the underground system must be installed in the heated area.
      ii. Pressurized air supply capable of pressurizing the system in thirty minutes and less.
      iii. An accelerator to be used when the system capacity exceeds 500 gallons.
      iv. An anti-flood device to prevent water from entering the accelerator when the system trips.
      v. A water motor alarm or an electric alarm.
      vi. Valve trimmings and pressure gages.
   d. Fire department connection and check valve with ball drip, if it is not provided in an underground system.
   e. A system of piping which carries water from the underground source to the sprinklers. The pipe size is determined by hydraulic calculations.
   f. Sprinkler type, temperature rating, orifice size, and coverage is determined by the design criteria, or recommended by the authorities having jurisdiction.
   g. Hangers and earthquake bracing are required. The type and spacing is determined by the design criteria or recommended by the authorities having jurisdiction.
   h. An inspector test connection is required on each system. The inspector test connection simulates the flow of water from one sprinkler head to assure the sprinkler system alarm will work.

System Operation

In case of a fire, heat fuses the sprinklers and that causes air in the pipe to escape. The dry pipe valve opens and allows the flow of water through the system and sounds the alarm. The alarm will continue to sound as long as water flows in the system, and the water flows until it is shut off. A dry pipe valve with an accelerator will trip faster and at a lower air pressure.

5 Exposure Protection System

Protection of Buildings from Exposure Fire

The spread of fire from one building to another across a vacant space is of great concern to building owners, insurance companies, and fire departments. The spread of fire may be due to connective and radiative heat transfer, or flying burning debris.

There are several means of fire protection that can be employed to protect the building from exposure fires, some of which are:

a. Total automatic sprinkler protection.
b. Clear space between buildings.
c. Walls of noncombustible materials between the building and its exposure.
d. Fire walls between the building and its exposure.
e. Extension of the exterior walls to form prapets or wings.
f. Automatic outside water curtain for combustible walls.
g. Elimination of openings.
h. Glass block panels in openings.
i. Wired glass in steel sash.
j. Automatic or deluge sprinkler system over outside openings
k. Automatic fire shutters on openings.
l. Automatic fire doors on door openings.
m. Automatic fire dampers on wall openings.

Components of the System

1. Water Supply
   An adequate water supply taken from the city main, gravity tank, pressure tank, storage tank, and well.
2. Underground System
   a. Piping – Cast iron, ductile iron, cement, or plastic.
   b. Control valves – Post Indicator Valve (PIV).

c. Valve Pit – Two Outside Stem & Yoke (OS & Y) gate valves, one Check Valve (CV) or detector check valve, a Fire Department Connection (FDC), a check valve with ball drip, a backflow preventer, a full flow meter, or a combination or equipment may be locally required.

d. Fire hydrants – Consult with the local fire department for location and thread requirements.

3. Overhead System

a. Wall PIV, OS & Y gate valves, or other type of control valves, if it is not provided in the underground system.

b. Alarm valve with a water motor alarm.

c. Check valve if a water flow alarm is used.

d. Fire department connection and check valve with a ball drip, if it is not provided in the underground system.

e. Waterflow detector can be installed in place of an alarm valve.

f. A system of piping which carries water from an underground source to the sprinklers. The pipe size is determined by hydraulic calculations.

g. Sprinkler type, temperature rating, orifice size, and coverage is determined by the design criteria, or recommended by the authorities having jurisdiction.

h. Hangers and earthquake bracing are required. The type and spacing is determined by the design criteria, or recommended by the authorities having jurisdiction.

i. An inspector test connection is required on each system. The inspector test connection simulates the flow of water from one sprinkler head to assure the sprinkler system alarm will work.

System Operation

In case of a fire, the heat fuses sprinklers and that causes the flow of water. This would fill the alarm line which closes the pressure switch and sound the alarm.

If a waterflow detector were used in the system, it would also activated by the flow of water. The water would force the paddle inside the pipe up, activating the tine delay mechanism. This action would set off an electric alarm.

All alarms would continue to sound as long as water flows in the system, and the water flows until it is shut off.

Firecycle System

Introduction

A firecycle system is a fixed fire protection system which utilizes water as an extinguishing agent. It also employs sprinkler system, heat detectors, and electrical controls. A firecycle system is a time delayed, preaction type, recycling system which automatically turns the water on when the temperature rises above the detector operating temperature and turns it off when the temperature falls below it. It is commonly used where water damage must be kept to a minimum.

The chance for accidental water damage is eliminated, because water can only be turned on by the detection system. There is no need to turn the water off for maintenance work or after a fire because the system piping is filled with pressurised air or gas.

Components of the system

1. The Design Problems
   a. Type of hazard.
   In order to apply the appropriate codes and standards, specific information about the hazard of the area to be protected must be obtained.
   b. System's purpose.
   System should be designed to extinguish fire in places where water damage must be kept to a minimum.
   c. Area to be protected.
   The maximum number of sprinklers allowed to be controlled by a Firecycle System is 1,000.
   d. Water supply.
   Water demand must be hydraulically calculated and an adequate water supply taken from the city main or existing system.
   e. Drainage.
   A preaction system delivers a very large volume of water to the fire area. The building owner must be notified of the amount of water his drain system has to drain, so he

can take appropriate measures to insure that his drain system has the capacity to drain the water.

f. Combustible liquids.

If there is floatable combustible liquid in the protected area, an adequate drainage or diking system should be installed to insure that a fire will not be transferred to adjacent areas.

2. Underground System
   a. Piping – Cast iron, ductile iron, cement, or plastic.
   b. Control valves – Post Indicator Valve (PIV).
   c. Valve Pit – Two Outside Stem & Yoke (OS & Y) gate valves, one Check Valve (CV) or detector check valve, a fire department connection (FDC), a check valve with ball drip, a backflow preventer, a full flow meter, or a combination of equipment may be locally required.
   d. Fire hydrants – Consult with the local fire department for location and thread requirements.
3. Overhead System
   a. A control valve, if it is not provided in the underground system.
   b. A check valve, if it is not provided in the underground system.
   c. Firecycle control panel.
   d. Firecycle trim box.
   e. System piping.
   f. Batteries and battery charger.
   g. Water motor alarm.
   h. Sprinklers.
   i. Firecycle detectors and cables.
   j. Main drain valve and intermediate drain valve.
   k. Priming valve.
   l. Position indicator.
   m. Strainer.
   n. Electric alarm bell.
   o. Maintenance air.
   p. Pressure switch
   q. Alarm test valve
   r. Alarm shut-off valve

System Operation

When there is a fire, the nearest detector is heated to its trip point. That causes the detector circuit to open, and relays in the control panel shut-off the power to the solenoid valves and sound the fire alarm bell. When the solenoid valves open, water is released and the system fills up. Now the sprinkler system is ready for operation. When a sprinkler reaches the operating temperature, it opens and sprays the fire area. Then the fire area cools and the temperature falls below the trip point. This causes the detector circuit to close and energize the timer. Sprinklers continue to operate from one to five minutes, depending upon the timer setting. When the timer cycle is complete, the water is turned off and the sprinklers stop spraying. This cycle will continue for as long as the heat is above the trip point.

7 Foam System

The foam system as a means of fire protection has greatly improved since its inception in 1930s. A foam system is designed to form a floating blanket over flammable liquids and prevent or extinguish fire by denying it access to oxygen. It also stops reignition by preventing the formation of combustible vapors. In addition, it provides exposure protection from adjacent fires.

There are three basic foam systems:

a. Protein-foam concentrates: This foam consists of products from protein hydrolysis, stabilizing additives, and inhibitions. The recommended concentration for this foam is three or six percent.

b. Fluoroprotein-foam concentrates: This foam is similar to protein-foam concentrates except that it has synthetic, fluorinated surfactant additives. It has an air excluding foam blanket. It also has an inhibiting agent to prevent vapor combustion. The recommended concentration of this foam is also three and six percent.

c. Aqueous Film Forming Foam Concentrates (AFFF): This foam consists of a fluorinated surfactant with stabilizers and additives. It excludes air and develops a surface film to prevent vapor combustion. The recommended concentration of this foam is also three and six percent.

There are three basic discharge devices that are installed in foam systems:

a. Foam-water sprinklers: This type of sprinkler has an open air-aspirating head designed especially for foam systems.

   It has an open barrel foam maker at the bottom, leading to the deflector at the top. The deflector discharges foam in a circular motion, the same pattern as there is with the regular sprinklers.

b. Foam-water spray nozzles: This type of nozzle is also an air-aspirating device which discharges foam or water in a directional pattern.

c. Standard sprinklers: Regular upright or pendent standard sprinklers have a circular pattern of fifteen feet in diameter.

1. The Design Problems
   a. Type of hazard.
      In order to apply the appropriate codes and standards, the specific information about the hazard of the area to be protected must be obtained.
   b. System's purpose.
      System should be designed to extinguish and/or prevent fire in places where a flammable liquid is present or there is a chance that it might leak, such as in an aircraft hangar.
   c. Area to be protected.
      Hydraulic calculation will determine the size of the system needed to protect the fire area.
   d. Water supply.
      Hater demand must be hydraulically calculated and an adequate water supply must be available to supply the system for a minimum of sixty minutes.
   e. Drainage.
      A foam water deluge system delivers a very large volume of water to the fire area. The building owner must be notified of the amount of water his drain system has to drain, so he can take appropriate measures to insure that his drain system has the capacity to drain the water.
   f. Combustible liquids.
      If there is a floatable combustible liquid in the protected area, an adequate drainage or diking system should be installed to insure that a fire will not be transferred to adjacent areas.
2. Underground System
   a. Piping – Cast iron, ductile iron, cement, or plastic.
   b. Control valves – Post Indicator Valve (PIV).
   c. Valve Pit – Two Outside Stem & Yoke (OS & Y) gate valves, one Check Valve (CV) or detector check valve, a Fire Department Connection (FDC), a check valve with ball drip, a backflow preventer, a full flow meter, or a combination of equipment may be locally required.
   d. Fire hydrants – Consult with the local fire department for location and thread requirements.
3. Overhead System
   a. A control valve, if it is not provided in the underground system.
   b. A check valve, if it is not provided in the underground system.
   c. Foam-water sprinklers.
   d. Foam-water spray nozzles.
   e. Standard sprinklers.

f. Air foam concentrates.
g. Air foam concentrates proportioning means.
h. Pumps.
i. Power supply.
j. Air foam concentrate storage tanks.
k. Pressure on air foam concentrate lines.
l. Alarms.
m. Strainers for water and air foam concentrates.
n. Water supply reservoir.
o. Foam agent reservoir.
p. Flow control valves.
q. Oscillating monitor.
r. Water-foam fire hoses.

System Operation

When there is a fire, the detection system signals the release control panel. The release control panel in turn opens control valves supplying water and foam to the nozzles and sounds the alarm. Foam concentrate is proportionally mixed with water in the supply piping. The flow control valve operates as a deluge valve, as well as a pressure regulating valve to insure the correct proportion of water and foam concentrate.

8 Halon Systems

Halon 1301 Fire Protection System

Halon 1301 is a colorless, odorless, electrically non-conductive gas that has proven to be an effective fire extinguishing agent.

This type of system combines very effective detection systems along with specially developed devices for high speed discharge. The system will extinguish fire by a chemical breakdown of the combustion process.

Halon 1301 systems are electrically operated and have a design discharge of less than five seconds. Agent storage containers are strategically located in the fire area, allowing high speed suppression of fire and keeping damage to a minimum.

These systems are generally recommended to protect electronic equipment, such as computers. They have the added advantage of causing less refrigeration of the air within the protected area than a CO_2 discharge. This is important because in some cases the electronic equipments are adversely effected by the sudden drop in temperature.

Components of the System

1. Agent Storage Containers
   Agent containers are designed to hold from 6 to 196 pounds of Halon 1301.
2. Control Unit
   Upon the operation of detectors, the control unit provides power for agent storage containers, causing a local alarm to actuate. Up to twenty agent storage containers can be controlled by one control unit.
3. Battery Unit
   The battery standby unit is designed to provide adequate energy to the system in case of a power failure.
4. Interface Firing Module
   The Interface Firing Module provides power and signal interface necessary to expand a fire alarm system to an automatic fire detection and extinguishing system.

5. Main/Reserve Switch
   The main/reserve switch is used to switch to either main or reserve agent containers.
6. Supervisory Station
   The supervisory station is designed for use in the systems which do not use a control unit.
7. Manual Pull Station
   The manual pull station is designed to electrically operate agent storage containers from a remote location by manual operation.
8. Initiator
   The initiator is an electro-explosive device used to release the extinguishing agent from an extinguisher.
9. Nozzle
   The nozzle is designed to discharge liquid Halon 1301 in a 120 degree discharge cone.
10. Thermal Detector
    Thermal detectors activate alarms and actuate the extinguishing system.
11. Rotating Beacon
    The rotating beacon is used where a visual indication is desired.
12. Alarm Light
    The alarm light is used where a visual indication is desired.
13. Product of Combustion Detector
    This unit is a smoke detector with an alarm response of about five seconds.

System Operation

When the system is activated, either by detectors or by manual operation, the connected initiators will fire, resulting in the discharge of the extinguishing agent from its containers. Alarms will be activated and any interlocked operating equipment will be shut down.

9 Preaction System

Introduction

A preaction system is a fixed fire protection system which is similar to a deluge system except that it employs closed sprinklers in the system piping and a detection system. The detection system can be hydraulic, pneumatic, electric, or manual. Detection systems are designed to detect the heat and sound alarm before sprinklers go off. Preaction systems are usually supervised to monitor the system.

Types of Preaction System

There are several types of preaction system. The basic three types are regular, single interlocked, and double interlocked.

a. Regular or non-interlock preaction system

   This system utilizes a deluge or dry valve which can be opened by the fusing of a sprinkler or by the operation of the detection system. The sprinkler system piping contains pressurized air or gas. The system is supervised to indicate low pressure only.

   The objective of this system is to fill the piping system with water before the fusing of sprinklers, giving the sprinklers a better chance at fighting the fire.

   Non-interlock preaction system is mostly used in a large unheated area where it exceeds the capacity of a dry valve system.

b. Single interlocked preaction system.

   The actuation of a single detection system trips the deluge valve and fills the system piping with water before the fusing of sprinkler heads. This allows sprinklers to be ready to fight fire at the tine of fusion.

c. Double interlocked preaction system.

   In a double interlocked system, as the title suggests, there are two safety systems against the accidental filling of system piping with water. First the detection system must actuate, then the sprinklers must fuse before water is allowed to fill the system. If only one of then is activated, the alarm sounds but water will not fill the system. This type of system frequently used in freezers or refrigerated areas.

Components of the System

1. The design problems
   a. Type of hazard.
      In order to apply the appropriate codes and standards, the specific information about the hazard of the area to be protected must be obtained.
   b. System's purpose.
      A system should be designed to extinguish fire in freezers and refrigerated areas, and also where a dry valve is not adequate to protect a large area.
   c. Area to be protected.
      The maximum number of sprinklers allowed to be controlled by a preaction system is 1,000.
   d. Hater supply.
      Water demand must be hydraulically calculated and adequate water supply taken from city main, or existing system.
   e. Drainage.
      A preaction system delivers a very large volume of water to the fire area. The building owner must be notified of the amount of water his drain system has to drain, so he can take appropriate measures to insure that his drain system has the capacity to drain the water.
   f. Combustible liquids.
      If there is floatable combustible liquid in the protected area, an adequate drainage or diking system should be installed to insure that a fire will not be transferred to adjacent areas.
2. Underground System.
   a. Piping – Cast iron, ductile iron, cement, or plastic.
   b. Control valves – Post Indicator Valve (PIV).
   c. Valve pit – Two Outside Stem & Yoke (OS & Y) gate valves, one Check Valve (CV) or detector check valve, a Fire Department Connection (FDC), a check valve with ball drip, a back- flow presenter, a full flow meter, or a combination of equipment may be locally required.
   d. Fire hydrants – Consult with the local fire department for location and thread requirements.
3. Overhead System
   a. Hall PIV, OS & Y gate valves, or other type of control valves, if they are not provided in the underground system.
   b. A check valve, if it is not provided in the underground system.
   c. Deluge or dry valve with all the trimming and release equipment.

d. Emergency release.
e. System piping.
f. Release system piping.
g. Water motor alarm.
h. Sprinklers.
i. Thermostatic release.

System Operation

When a deluge valve is used, the system operates the same as a deluge system would. If a dry valve is used, the system operates the same as a dry pipe system would.

10 Wet Pipe System

Wet Pipe Sprinkler System

A wet pipe sprinkler system is a fixed fire protection system using piping filled with pressurized water. The system is activated by fusible sprinkler heads, which go off at a preset temperature, spray the fire area, and put out the fire or hold it in check until Firemen arrive.

A wet pipe sprinkler system can be installed in any heated building which is structurally sound to support system piping filled with water. The low cost of equipment, installation, and maintenance plus the speed of operation account for the fact that 75 per cent of all the fire sprinkler systems in use today are of the wet pipe type.

The maximum area a wet pipe system can protect is 52,000 square feet per floor. For example, a building with several stories can be protected with one wet pipe system, provided the area of each floor does not exceed 52,000 square feet.

Components of the System

1. Water Supply
   An adequate water supply taken from city main, gravity tank, pressure tank, storage tank, and well.
2. Underground System
   a. Piping – Cast iron, ductile iron, cement, or plastic.
   b. Control valves – Post Indicator Valve (PIV).
   c. Valve pit – Two Outside Stem & Yoke (OS & Y) gate valves, one Check Valve (CV) or detector check valve, a fire department connection (FDC), a check valve with ball drip, a back-flow preventer, a full flow meter, or a combination of equipment may be locally required.
   d. Fire hydrants – Consult with the local fire department for location and thread requirements.
3. Overhead System
   a. Hall PIV, OS & Y gate valves, or other type of control valves, if there are not provided in the underground system.

b. Alarm valve with water motor alarm.
c. Check valve if a water flow alarm is used.
d. Fire department connection and check valve with ball drip, if there are not provided in the underground system.
e. Waterflow detector can be installed in place of an alarm valve.
f. A system of piping which carries water from an underground source to the sprinklers. The pipe size is determined by hydraulic calculations.
g. Sprinkler type, temperature rating, orifice size, and coverage is determined by the design criteria or recommended by the authorities having jurisdiction.
h. Hangers and earthquake bracing are required. The type and spacing is determined by the design criteria or recommended by the authorities having jurisdiction.
i. An inspector test connection is required on each system. The inspector test connection simulates the flow of water from one sprinkler head to assure the sprinkler system alarm will work.

System Operation

In case of a fire, the heat will fuse sprinklers and that causes the flow of water. The alarm valve clapper will open and allow water to fill the retarding chamber. This would fill the alarm line which closes the pressure switch and sound the alarm.

If a waterflow detector were used in the system, it would also become activated by the flow of water. The water would force the paddle inside the pipe up, activating the time delay mechanism. This action would set off an electric alarm. All alarms would continue to sound as long as water flows in the system, and the water would flow until it is shut off.

11 Explosion Protection System

Introduction

An explosion protection system is designed for installation in an area where a definite explosion hazard exists. This system has the unique capability to prevent an explosion after it starts and before it has a chance to do major damage.

This system operates on the principle that there is a short but measurable time between the ignition and the explosion. Although this delay time is only a few milliseconds, it is enough time for the system to react and prevent an explosion. An explosion protection system consists of a power unit, detectors, suppressors, and extinguishers.

Components of the System

1. Power Unit
   It converts different signals from detectors into milliseconds current pulse which actuates explosion suppression devices. It supervises all wiring circuits and has ground indicator, safety switch, and back up batteries.
2. Pressure Detectors
   Highly sensitive. Ion inertia, diaphragm type design provides sensitivity to the rate of rise and/or static pressure pulse.
3. Hemispherical Suppressors
   The explosive actuators located at the center is fired by an electric charge from the control unit. Hydrostatic pressure ruptures the hemisphere and disperses the suppressing agent at a velocity of 200 feet per second.
4. High Rate Discharge Extinguisher
   It is used where a high volume of suppressant is required. Hydrostatic pressure ruptures the burst disc and extinguishing agents are discharged. This process is completed in a few milliseconds.

12 Light Hazard Occupancies

Light Hazard

Thin type of occupancies involves a fire area in which the quantity and quality of combustible materials are low.

Some of the light hazard buildings are:

- Apartments
- Churches
- Clubs
- Hospitals
- Libraries
- Museums
- Nursing Homes
- Offices
- Restaurants
- Schools
- Theaters

Spacing

1. In an area with a noncombustible smooth ceiling and a concrete and steel construction, the maximum area per sprinkler is 225 sq. ft. (20.9 m^2). With other noncombustible ceilings, the maximum area per sprinkler is 196 sq. ft. (18.2 m^2).
2. In an area with a combustible smooth ceiling, sheathed with plasterboard, metal, or wood lath construction, the maximum area per sprinkler is 168 sq. ft. (15.6 m^2).
3. In an area with an exposed wood joist ceiling, the maximum area per sprinkler is 130 sq. ft. (12.1 m^2).
4. In an area with fiberboard or other combustible exposed ceilings, the maximum area per sprinkler is 120 sq. ft. (11.1 m^2).
5. Sidewall sprinklers are allowed to be installed in light hazard occupancies. The fire area must have a smooth ceiling or bays. The bays must have a maximum width of 30 ft. (9 m).

6. The maximum distance between sprinklers, either on branch lines or between branch lines, is 15 ft.
7. The maximum distance between hangers on a 1½ inches (38 mm) steel pipe or larger is 15 ft. (4.5 m). For a steel pipe smaller than a 1½ inches (38 mm), the maximum distance between hangers is 12 ft. (3.6 m).
8. Earthquake bracing may be installed on a 2½ inches (64 mm) steel pipe and larger. The installation shall be on each turn and every 40 ft. (12 m) maximum on a lateral run.

Design

There are two methods of pipe nixing for light hazard occupancies.

1. Light hazard pipe schedules are used when hydraulic calculation is not required.
2. Hydraulically calculated systems calculate pipe sizing as well as pressure and flow. Hydraulic calculation criteria are determined from a NFPA chart, or they are provided by the authorities having jurisdiction.

13 Ordinary Hazard Occupancies

Ordinary Hazard

This type of occupancy involves a fire area in which the quantity and quality of combustible materials are moderate, and fires with moderate rates of heat release are expected. Ordinary hazards are divided into three groups, based on their combustibility and their heat release:

1. Ordinary Hazard Group I
   In this group usually the rates of combustibility are low, quantity of combustibles is moderate, stock piles of combustibles do not exceed 8 ft. (2.4 m), and the rates of their heat release are moderate.

Some of the ordinary hazard group I are:

- Automobile parking garages.
- Bakeries
- Beverage manufacturing plants.
- Canneries
- Dairy products manufacturing and processing plants
- Electronic plans
- Glass product manufacturing plants
- Laundries
- Restaurant service areas

2. Ordinary Hazard Group II

In this group quantity and combustibility are moderate, stock piles do not exceed 12 ft. (3.7 m), and the rates of their heat release are moderate.

Some of the ordinary hazard group II are:

- Cereal mills
- Chemical plants
- Cold storage warehouses
- Confectionery products
- Distilleries
- Library stock rooms
- Mercantile centers

- Machine shops
- Metal working shops
- Printing and publishing centers
- Textile manufacturing plants
- Tobacco product manufacturing centers
- Mood product assembly shops

3. Ordinary Hazard Group III

In this group quantity and combustibility are high, and the rates of their heat release is also high.

Some of the ordinary hazard group III are:

- Feed mills
- Paper mills
- Paper processing plant
- Piers and wharves
- Repair garages
- Tire manufacturing plant
- Warehouses containing paper products, household furniture, paints, general storage, and whiskey
- Wood machines.

Spacing

1. In an area with a noncombustible smooth ceiling, the maximum area per sprinkler is 100 sq. ft. (9.3 m^2).
2. In an area with a combustible ceiling, sheathed with plasterboard, metal, wood lath and plaster, and a wood fiberboard construction, the maximum area per sprinkler is 80 sq. ft. (7.4 m^2).
3. In all other areas, the maximum area per sprinkler is 130 sq. ft. (12.1 m^2).
4. The maximum distance between sprinklers, either on branch lines or between branch lines, is 15 ft.
5. The maximum distance between hangers on a 1½ inches (38 mm) steel pipe or larger is 15. ft (4.5 m). For a steel pipe smaller than a 1½ inches (38 mm), the distance between hangers is 12 ft. (3.6 m).
6. Earthquake bracing may be installed on a 2½ inches (84 mm) or larger steel pipe. The installation shall be on each turn and every 40 ft. (12 m) maximum on a lateral run.

Design

There are two methods of pipe sizing for ordinary hazard occupancies.

1. Ordinary hazard pipe schedules are used when hydraulic calculation is not required.
2. Hydraulically calculated systems calculate pipe sizing as well as pressure and flow. Hydraulic calculation criteria are determined from a National Fire Protection Association (NFPA) chart, or they are provided by the authorities having jurisdiction.

14 Extra Hazard Occupancies

Extra Hazard

This type of occupancy involves a fire area in which the quantity and quality of combustible materials are very high, and fires with very high rates of heat release are expected. Extra hazards are divided into two groups, based on their combustibility and their heat release.

1. Extra Hazard Group I

   In this group usually the rates of combustibility are high and the quantity of combustibles is high, but the occupancy does not contain very much flammable and combustible liquid.

   Some of the extra hazard group I are:

   - Machine shops using hydraulic fluid
   - Plywood shops
   - Printing shops
   - Rubber manufacturing plants
   - Saw mills
   - Textile manufacturing plants
   - Upholstering with plastic foams

2. Extra Hazard Group II

   In this group quantity and combustibility are very high, with substantial amounts of flammable and combustible liquids.

   Some of the extra hazard group II are:

   - Asphalt saturating
   - Flammable liquids
   - Flow coating
   - Mobile homes assemblies
   - Open oil quenching
   - Solvent cleaning
   - Varnish and paint dipping

Spacing

1. In all areas the maximum area per sprinkler is 100 sq. ft. (9.3 m^2), unless it is calculated.
2. The maximum distance between sprinklers, either on branch lines or between branch lines, is 12 ft.
3. The maximum distance between hangers on a 1½ inches (38 mm) steel pipe or larger is 15 ft. (4.5 m). For steel pipe smaller than a 1½ inches (38 mm), the maximum distance between hangers is 12 ft. (3.6 m).
4. Earthquake bracing may be installed on a 2½ inches (64 mm) or larger steel pipe. The installation shall be on each turn and every 40 ft. (12 m) maximum on a lateral run.

Design

There are two methods of pipe sizing for extra hazard occupancies:

1. Extra hazard pipe schedules are used when hydraulic calculation is not required.
2. Hydraulically calculated systems calculate pipe sizing as well as pressure and flow. Hydraulic calculation criteria are determined from a NFPA chart, or there are provided by the authorities having jurisdiction.

15 Introduction to Hydraulic Calculations

Hydraulically Designed Fire Protection Systems

In a hydraulically designed system, the pipe sizes are selected based on pressure loss for a given density. The pressure is in psi and the density is in gpm. This provides uniformity of pipe sizes over the calculated area as well as over the entire floor area. This type of system also allows the selection of pipe sizes in accordance with the water supply available. In addition, we must consider that the density and calculated area will change in accordance with the occupancy hazard.

The design basis for such a system supersedes schedule pipe sizing, but spacing and other applicable rules shall be observed.

The design criteria for a hydraulically calculated system shall be obtained from the authorities having jurisdiction. In the absence of that, the information must be obtained from the National Fire Protection Association graph.

The following information must be on the drawing before submitting it to the authorities having jurisdiction for their approval.

1. System requirement
   a. Area of calculation.
   b. Density.
   c. Area per sprinkler.
   d. Allowance for inside hose stations.
   e. Allowance for outside fire hydrants.
   f. Allowance for in-rack sprinklers.
2. Water Supply information
   a. Location and elevation of the nearest tested fire hydrant.
   b. Static pressure.
   c. Residual pressure.
   d. Flow.
   e. Date.
   f. Time.

g. Test conducted by or information supplied by...

3. Hydraulic reference points
4. Description and degree of sprinkler heads

Hydraulic Calculation Sheets

1. Hydraulic calculation cover sheets must contain the following information:
   a. Contract number.
   b. Client's name.
   c. Building name and address.
   d. Description of hazard.
   e. Authorities having jurisdiction.
   f. Name and location of engineer doing calculations.
   g. Date of calculation.
   h. System number.
   i. Sheet number.
   j. Drawing number.
   k. Basic design data:
      i. Density gpm/ft^2.
      ii. Calculated area ft^2.
      iii. Allowance for inside hose stations gpm.
      iv. Other water allowance gpm.
      v. Remarks.
   l. Flow test data:
      i. Static psi.
      ii. Residual psi.
      iii. GPM flowing.
      iv. Test gauge location and elevation.
      v. Flow hydrant location.
      vi. Remarks.
   m. Fire pump data:
      i. Rated gpm.
      ii. Rated psi.
      iii. Elevation.
      iv. Type of pump.
      v. Remarks.
   n. Elevated tank data:
      i. Discharge pipe size,
      ii. Tank bottom elevation.

o. Other data:
   i. Underground pipe.
   ii. Building height.

p. Calculation summary:
   i. GPM required.
   ii. PSI required.
   iii. Location.
   iv. Remarks.
   v. Abbreviations.

2. Hydraulic calculation sheet must contain the following information:
   a. Sheet number.
   b. Sprinkler description and location.
   c. K-factor of sprinkler heads.
   d. Hydraulic reference points.
   e. Flow in gpm.
   f. Friction loss in psi per foot of pipe.
   g. Friction loss in psi between reference points.
   h. Pipe sizes.
   i. Pressure in psi at each reference point.
   j. Velocity pressure and normal pressure between reference points.
   k. Pipe lengths.
   l. Equivalent pipe lengths for pipes, fittings and devices.
   m. In-rack sprinkler demand in gpm.
   n. Inside hose-rack demand in gpm.
   o. Contract name and contract number.

3. Hydraulic calculation graph sheet.
   This graph sheet is a semi-logarithmic graph on paper which horizontally shows pressure in psi and vertically shows flow in gpm. System requirements such as water supply, inside hose stations, and in-rack sprinkler demands are plotted on it.

Hydraulic Calculations

1. Friction loss formula
   Pipe friction loss calculations must be based on the Hazen-WIllians formula.

$$P = \frac{4.52\,Q^{1.85}}{C^{1.85} d^{4.87}}$$

P = Frictional resistance in psi per foot of pipe

Q = Flow in gpm

d = Internal diameter of pipe in inches
C = Friction loss coefficient

Metric system (SI Units):

$$Pm = 6.05 \times \frac{Qm^{1.85}}{C^{1.85} dm^{4.87}} \times 10^5$$

Pm = Frictional resistance in bar per meter of pipe
Qm = Flow in liter per minute
dm = Internal diameter of pipe in mm
C = Friction loss coefficient

2. Velocity pressure (Pv)
   The calculation of this velocity must be based on the formula

$$Pv = 0.001\,123\,Q^2/D^4$$

   Pv = Velocity pressure in psi
   Q = Flow in gpm
   D = Inside diameter of pipe in inches
   For conversion to the metric system (SI Units), use conversion tables.

3. Normal Pressure (Pn)
   The calculation of this velocity must be based on the formula

$$Pn = Pt - Pv$$

   Pt = Total pressure in psi
   Pv = Velocity pressure in psi

4. Hydraulic Junction Points
   Pressure at hydraulic junction points must be balanced within 0.5 psi. The higher pressure and the combined flow must be carried into the next hydraulic junction point for continued calculations.

5. Equivalent Pipe Lengths of Valves and Fittings
   The table provided at the end of this chapter must be to determine the equivalent lengths of pipe for fitting and devices.

6. Value of "C"
   The table provided at the end of this chapter must be used to determine the value of the "C" factor.

Calculation Procedure

In every system the calculated area must be the highest and the most remote area from the riser. The area, if possible, must be rectangular in shape with the long side parallel to the branch lines.

16 Hydraulically Designed Loop and Grid Systems

Introduction

There are basically three types of hydraulically designed systems. They are standard, loop, and grid systems.

Standard system

In a standard system, the end sprinkler is usually on the last piece of pipe at the end of the branch line. Branch lines are usually connected by a nipple to the cross main. Cross mains are connected to a feed main, and feed mains are connected to the system riser. Risers are connected to the underground fire line, and underground lines are connected to the water supply.

Consider the attached drawing. The spacing is 10′ × 13′=130 ft.

The number of sprinklers to be calculated = 1500/130 = 11.54.

The number of sprinklers to be calculated = 12 sprinklers.

The number of sprinklers to be calculated on each line = $\frac{1.2\sqrt{1500}}{13} = 3.58$ or 4 sprinklers on each line.

D = 0.16 gpm/ft².

K = 5.6

Q = 0.16 gpm/ft² x 130 ft² = 20.8 gpm

Q = $K\sqrt{P}$

$$P = \frac{Q^2}{K^2} = \frac{(20.8)^2}{(5.6)^2} = 13.8 \text{ psi}$$

Now follow steps 1 through 11 on a hand calculation sheet, and study the procedure of computer calculations for the same system.

Loop System

A loop system is a closed circuit of piping with at least two paths for water to get to the target area. Loop systems are used to reduce static pressure, therefore reducing pipe sizing.

The hydraulic calculations for loop systems are similar to standard systems, except that in loop systems water goes through at least two paths to get to the calculated area. Therefore, almost half of the water goes through half of the loop, and the remaining water goes through the other half of the loop.

This means that after the line calculations are done, the water splits almost in half, one half going through one leg of the loop and the other half going through the other leg. These two halves are balanced at the point where the feed main is connected to the loop. The rest of the calculations are exactly like standard calculations.

Grid System

A grid system is similar to a loop system except that in a loop system about half the lines are outside of the loop, but in a grid system all the lines are inside of the loop.

17 Fire Protection for Homes and Mobile Homes

Introduction

The purpose of this chapter is to provide a fire sprinkler system, in addition to smoke detectors, to protect families and their homes. In order to have maximum protection, sprinklers must be installed in all enclosed spaces and overhangs wider than four feet. However, the standard practice is to install sprinklers in all the rooms, kitchen, storage areas, and halls.

Authorities having jurisdiction must be consulted for compliance with codes and standards prior to the start of design and installation.

This type of sprinklers system is usually a wet pipe sprinkler system using a fixed piping filled with pressurized water. The system is activated by fusible sprinkler heads, which go off at a pre-set temperature, spray the fire area, and put out the fire or hold it in check until firemen arrive.

A wet pipe sprinkler system can be installed in any heated building which is structurally sound to support system piping filled with water.

Components of the System

1. Water Supply
   a. An adequate water supply taken from the city main, gravity tank, pressure tank, storage tank, and well.
2. Underground System
   a. Use the existing underground water main coming to the home.
3. Overhead System
   a. Control valve.
   b. Check valve.
   c. Waterflow detector.
   d. Pressure gauge.
   e. A ½ inch drain.
   f. A system of piping which carries water from an underground source to the sprinklers. The pipe size is usually one inch.

g. Sprinkler type, temperature rating, orifice size, and coverage is determined by the design criteria or recommended by the authorities having jurisdiction.

h. Hangers are required. The type and spacing is determined by the design criteria or recommended by the authorities having jurisdiction.

i. A ½ inch test connection.

System Operation

In case of a fire, the heat will fuse sprinklers and that causes the flow of water. This activates the waterflow detector and forces the paddle inside the pipe up, activating time delay mechanism. This action sets off an electric alarm. All alarms continue to sound as long as water flows in the system, and the water flows until it is shut off.

Design Criteria

1. The minimum flow for a single sprinkler is 18 gallons per minute. In an area with more than one sprinkler operating, the minimum flow per sprinkler shall be 13 gallons per minute.
2. The maximum spacing per sprinkler shall not exceed 144 sq. ft.
3. The maximum distance between sprinklers shall not exceed 12 ft.
4. The maximum distance between sprinklers and a wall or partitions shall not exceed 6 ft.
5. The minimum distance between sprinklers shall be 8 ft.
6. All sprinklers shall be of residential types, UL listed, and approved for installation by authorities having jurisdiction.
7. For deletion of sprinklers in snail spaces, consult authorities having jurisdiction.

18 Fire Protection for Mid-Rise Buildings

Definition

A mid-rise building in between 55 to 75 feet in height. Measurement shall be from the topmost occupied space to the lowest fire apparatus access road level or the building access, whichever is lower.

Exceptions

Buildings such as open parking garages, power plants, lookout towers, grain elevators, and similar structures with noncontinuous human occupancy, may be excluded by the authorities having jurisdiction.

Building Access

An exterior door which meets the following conditions.

1. Opening onto or adjacent to a public way that is accessible to emergency vehicles, such as fire and police departments.
2. Located not more than two feet above the adjacent ground level.
3. Leading to a space, hall, room, or area which leads to the rest of the building.

Fire Alarm Systems

Every mid-rise building shall be provided with an approved electrically supervised fire alarm system conforming to the Uniform Fire Code, National Electrical Code and applicable codes and standards set forth by the authorities having jurisdiction.

1. Local Alarm: Upon activation of the automatic fire sprinkler system, an automatic alarm signal will sound.
   The local alarm may employ any sounding devices which are approved by the authorities having jurisdiction. The sounding devices of such a system shall have a distinctive tone and shall be arranged to emit intermittent, prolonged or continuous sound signals for a full period of 10 seconds, to be immediately followed by an intermission or period of silence of 5 seconds, before the signal is repeated. Such a signal shall continue to sound until manually terminated at a Lobby Fire Alarm Control Panel.

2. Level of Alarm: The level of audibility shall not be less than 10 db above ambient noise levels measured four feet above the floor, unless otherwise recommended by the authorities having jurisdiction.
3. Supervision: Fire alarm and trouble signals shall be automatically transmitted to one of the following:
   a. An approved central station.
   b. The supervisory station.
   c. The local fire department, when approved.
4. Annunciation: Alarm, waterflow, and trouble signals shall be annunciated at the Lobby Fire Alarm Control Panel by means of an audible signal and a visual display, which indicates the building, floor, zone, or other designated area from which the alarm, waterflow, or trouble signal originated. Zoning shall be done in accordance with the following:
   a. When the system serves more than one building, each building is considered as a separate zone.
   b. In a aid-rise building, each floor is considered as a separate zone.
   c. Each section of floor separated by a fire wall or a separation wall is considered as a separate zone.
   d. As it has been recommended by the authorities having jurisdiction.

Automatic Fire Detection Systems

Every mid-rise building shall be protected with an approved electrically supervised automatic fire detection system. This system shall have automatic detectors which will respond to visible or invisible particles of combustion in the following location:

1. In every mechanical room.
2. In the main circulating air duct on the downstream side of the filters.
3. In each room or space served by a return air duct.
4. At every opening through any assembly having a fire- resistive rating of two hours or more.

Automatic Fire Sprinkler Systems and Standpipes

Every mid-rise building shall be provided with an approved combination standpipe system. A combination standpipe system provides two services at the same time.

1. Provides the required 2½ inch outlet at each floor for fire department use.
2. Provides the required supply outlet at each floor for the automatic fire sprinkler system.

The combination standpipe shall he provided with an approved fire pulping system capable of delivering a minimum of 500 gpm for the first standpipe plus 250 gpm for each additional standpipe. The maximum water supply needed shall not exceed 1,500 gpm.

The pumping system shall deliver the required supply at the topmost outlets, while maintaining a residual pressure of 65 psi at the roof outlet of each standpipe.

The city water main may be used as a water supply, and an on-site water supply may not be required.

For water supply and pumping system requirements, consult the authorities having jurisdiction.

Fire Department Communication System

An approved electrically supervised sound powered system shall be provided in every mid-rise building for fire department use.

1. This system shall be of a type approved by the fire department. Such a system shall provide communication between the Lobby Fire Alarm Control Panel and remote facilities in the following locations:
   a. Elevators.
   b. Elevator lobbies.
   c. Every enclosed exit stairway at each floor level including the ground floor that leads directly to the exterior.

Ventilation

Natural or mechanical ventilation for the removal of smoke shall be provided in every floor of a mid-rise building.

The interior air circulation system may be used provided, under fire condition, it exhausts at a minimum of six air changes per hour and exhausts directly to the outside of the building without recirculating.

Other designs may be acceptable, if they meet the approval of the authorities having jurisdiction.

Elevators

1. General: In every bank of elevators in a mid-rise building, one elevator shall be provided for fire department use.
   This elevator must have access to each floor of the building.
2. Lobbies: All elevators shall open into a lobby on each floor except the main floor of the building entry.
   Lobbies shall be separated from corridors by a one hour fire door. These doors are designed to close automatically upon activation of a detector which will respond to the start of a fire.
3. Size of Cab
   Elevator cars assigned for fire department use shall have a recessed area or removable ceiling so a 9-foot ladder can be carried aboard. This elevator shall be large enough to house a 22 × 78 inch stretcher, and its door must be at least 42 inches wide.

Standby Power

1. Required Installation: An on-site standby power system shall be provided for every mid-rise building. In the event of failure of the normal source of electric service, the standby

power system shall provide an alternate source of electrical energy to serve at least the designated power loads. The system may consist of on-site generator, a system of batteries or both.

2. Power Loads: The power load requirements for sizing the standby power system shall include, but not be limited to the following:
   a. Exit signs and exit illumination.
   b. Fire alarm system.
   c. Fire detection system.
   d. Sprinkler alarm system.
   e. Elevator assigned for fire department use.
   f. Fire pumps.
   g. Lighting circuits supplying elevators, elevator lobbies, and the generator room.
3. Transfer Time: A standby power generator shall be equipped with suitable means for automatically starting the generator upon failure of the normal electrical service. It shall provide for the automatic transfer and operation of electrical systems at full power within 60 seconds of normal service failure.
4. Fuel Supplies: On-site fuel supplies for a standby power generator shall be sufficient for at least 8 hours of full demand operation.

Emergency Electrical Systems

1. Required Installation: Electrical systems and equipment shall be installed in every mid-rise building in accordance with the provisions of the National Electrical Codes.
   a. Exit signs and exit illumination.
   b. Fire alarm system.
   c. Fire detection system.
   d. Sprinkler alarm system.
2. Transfer Time: The emergency power supply system shall be so designed that upon failure of the normal electrical service, the emergency electrical systems and equipment shall be automatically transferred and operated within 10 seconds of such normal service failure.

Locking of Stairway Doors

All stairway doors which must be locked from the stairway side shall have the capability of being unlocked without unlatching upon a signal from the Lobby Fire Control Panel. Upon failure of electrical power, the locking mechanism shall be retracted to the unlocked position.

19 Fire Protection for High-Rise Buildings

Definition

A high-rise building is over 75 foot in height. Measurement shall be from the topmost occupied space to the lowest fire apparatus access road level or the building access, whichever is lower.

Exceptions

Buildings such as open parking garages, power plants, lookout towers, grain elevators, and similar structures with noncontinuous human occupancy may be excluded by the authorities having jurisdiction.

Building Access

An exterior door which meets the following conditions.

1. Opening onto or adjacent to a public way that is accessible to emergency vehicles, such as those of fire and police departments.
2. Located not more than two feet above the adjacent ground level.
3. Leading to a space, hall, room, or area which leads to the rest of the building.

Fire Alarm Systems

Every high-rise building shall be provided with an approved electrically supervised fire alarm system conforming to the Uniform Fire Code, National Electrical Code and applicable codes and standards set forth by the authorities having jurisdiction.

1. Local Alarm: Upon activation of the automatic fire sprinkler system, an automatic alarm signal will sound.
   The local alarm may employ any sounding devices which are approved by the authorities having jurisdiction. The sounding devices of such a system shall have a distinctive tone and shall be arranged to emit intermittent, prolonged or continuous sound signals for a full period of 10 seconds, to be immediately followed by an intermission or period of silence of 5 seconds, before the signal is repeated. Such a signal shall continue to sound until manually terminated at a Lobby Fire Alarm Control Panel.

2. Level of Alarm: The level of audibility shall not be less than 10 db above ambient noise levels measured four feet above the floor, unless otherwise recommended by the authorities having jurisdiction.
3. Supervision: Fire alarm and trouble signals shall be automatically transmitted to one of the following:
   a. An approved central station.
   b. The supervisory station
   c. The local fire department, when approved.
4. Annunciation: Alarm, waterflow, and trouble signals shall be annunciated at the Lobby Fire Alarm Control Panel by means of an audible signal and a visual display, which indicates the building, floor, zone, or other designated area from which the alarm, waterflow, or trouble signal originated. Zoning shall be done in accordance with the following:
   a. When the system serves more than one building, each building is considered as a separate zone.
   b. In a mid-rise building, each floor is considered as a separate zone.
   c. Each section of floor separated by a fire wall or a separation wall is considered as a separate zone.
   d. As it has been recommended by the authorities having jurisdiction.

Automatic Fire Detection Systems

Every mid-rise building shall be protected with an approved electrically supervised automatic fire detection system. This system shall have automatic detectors which will respond to visible or invisible particles of combustion in the following location:

1. In every mechanical room.
2. In the main circulating air duct on the downstream side of the filters.
3. In each room or space served by a return air duct.
4. At every opening through any assembly having a fire-resistive rating of two hours or more.

Automatic Fire Sprinkler Systems and Standpipes

Every mid-rise building shall be provided with an approved combination standpipe system. A combination standpipe system provides two services at the same time.

1. Provides a required 2 ½ inch outlet at each floor for fire department use.
2. Provides required supply outlet at each floor for the automatic fire sprinkler system.

The combination standpipe shall be provided with an approved fire pumping system capable of delivering a minimum of 500 gpm for the first standpipe plus 250 gpm for each additional standpipe. The maximum water supply needed shall not exceed 1,500 gpm.

The pumping system shall deliver the required supply at the topmost outlets, while maintaining a residual pressure of 65 psi at the roof outlet of each standpipe.

The city water main may be used as a water supply, and an on-site water supply may not be required.

For water supply and pumping system requirements, consult the authorities having jurisdiction.

Fire Department Communication System

An approved electrically supervised sound powered system shall be provided in every high-rise building for fire department use.

1. This system shall be of a type approved by the fire department. Such a system shall provide communication between the Lobby Fire Alarm Control Panel and remote facilities in the following locations:
    a. Elevator
    b. Elevator lobbies.
    c. Every enclosed exit stairway at such floor level including the ground floor that leads directly to the exterior.

Ventilation

Natural or mechanical ventilation for the removal of smoke shall be provided on every floor of a mid- rise building.

The interior air circulation system may be used provided, under fire condition, it exhausts at a minimum of six air changes per hour, and exhausts directly to the outside of the building without recirculating.

Other designs may be acceptable, if they meet the approval of the authorities having jurisdiction

Elevators

1. General: In every bank of elevators in a high-rise building one elevator shall be provided for fire department use. This elevator must have access to each floor of the building.
2. Lobbies: All elevators shall open onto a lobby on each floor except the main floor of the building entry.
   Lobbies shall be separated from corridors by a one hour fire door. These doors are designed to close automatically upon activation of a detector which will respond to the start of a fire.
3. Size of Cab: Elevator cars assigned for fire department use shall have a recessed area or removable ceiling so that a 9-foot ladder can be carried aboard. This elevator shall be large enough to house a 22 × 78 inch stretcher, and its door must be at least 42 inches wide.

Standby Power

1. Required Installation: An on-site standby power system shall be provided for every mid-rise building. In the event of failure of the normal source of electric service, the standby

power system shall provide an alternate source of electrical energy to serve at least the designated power loads. The system may consist of on-site generator, a system of batteries or both.

2. Power Loads: The power load requirements for sizing the standby power system shall include but not be limited to the following:
   a. Exit signs and exit illumination.
   b. Fire alarm system.
   c. Fire detection system.
   d. Sprinkler alarm system.
   e. Elevator assigned for fire department use.
   f. Fire pumps.
   g. Lighting circuits supplying elevators, elevator lobbies, and the generator room.
3. Transfer Time: A standby power generator shall be equipped with suitable means for automatically starting the generator upon failure of the normal electrical service. It shall provide for the automatic transfer and operation of electrical systems at full power within 60 seconds of normal service failure.
4. Fuel Supplies: On-site fuel supplies for a standby power generator shall be sufficient for at least 8 hours of full demand operation.

Emergency Electrical Systems

1. Required Installation: Electrical systems and equipments shall be installed in every high-rise building in accordance with the provisions of the National Electrical Codes.
   a. Exit signs and exit illumination.
   b. Fire alarm system.
   c. Fire detection system.
   d. Sprinkler alarm system.
2. Transfer Time: The emergency power supply system shall be so designed that upon failure of the normal electrical service, the emergency electrical systems and equipments shall be automatically transferred and operated within 10 seconds of such normal service failure.

Locking of Stairway Doors

All stairway doors which must be locked from the stairway side shall have the capability of being unlocked without unlatching upon a signal from the Lobby Fire Control Panel. Upon failure of electrical power, the locking mechanism shall be retracted to the unlocked position.

20 Fuel Use and Storage

Introduction

The arrangement of fuel within a given area has an impact on the fire hazard or fire suppression within that area.

In this chapter we will discuss fuel loading and arrangement.

Fuel loading is the amount of fuel in a given fire area. Fuel arrangement deals with the manner in which the fuel is stored.

The arrangement of the fuel can have a significant effect on the potential fire hazard, fire prevention, and fire protection.

Fuel Loading

Fire loading is the weight of combustible materials per square foot. By calculating the sum of the weight of all combustible contents and finish materials in a room and dividing the weight by the floor area of the room, an average fire loading factor is obtained. A combustible loading of less than 10 lb./ft. sq. is considered to be a low combustible loading; less than 25 lb.ft. sq. but greater than 10 lb./ft. sq. is considered to be a moderate combustible loading, and greater than 25 lb./ft. sq. is considered to be a high combustible loading.

Plastics and liquid hydrocarbons produce heat of combustion on a per pound basis which may exceed twice that of ordinary material such as wood or paper. Therefore, the categorization of fire severity based solely on the weight of combustible material over an area is not always correct. Furthermore, it must be recognized that the rate of heat release is influenced by several factors in addition to the total weight of fuel available for combustion.

Additional factors which affect the rate of fire within a compartment include:

1. The arrangement of materials within the enclosure.
2. The total calorific value of the combustible materials within the enclosure.
3. The form in which the combustible material is present.
4. The size and location of ventilation openings.
5. The relative geometry of the enclosure.
6. The insulating capacity of the enclosure boundaries.

The storage arrangements and safe practices differ for various materials. Combustible liquids, flammable gases, and combustible solids have different fire characteristics and therefore have different storage requirements for fire safety.

21 Flammable or Combustible Liquids

The fire hazards of a flammable or combustible liquid can also be minimized by controlling the environment. These controls can include controlling the temperature, oxygen concentration, and ensuring that the vapor concentration is outside the flammable range. These liquids should be stored and used at temperatures well below their flash point.

It is important to note that almost all flammable liquids produce heavier-than-air vapors which tend to settle at the floor level. Such vapors may flow along the floor or ground for long distances, be ignited at some remote point, and flash back. As a result, the provision for adequate ventilation at floor level near the vapor source is essential. Ventilation may be either by natural or artificial means, but mechanical ventilation is preferable to ensure that the airflow is sufficient under all conditions.

Spills of flammable liquids must be prevented whenever possible and must be controlled quickly so that the amount of liquid involved is minimized. To accomplish this, self-closing type valves should generally be provided on liquid dispensing lines. In addition, consideration should be given to installing overflow prevention devices on bulk storage tanks. These include flat valves, return lines, and specially designed pumps.

Often, the most practical manner to handle flammable liquids is to separate their area of storage from other areas. This can be as simple as providing a flammable liquid storage cabinet for small quantities of liquids or as elaborate as providing a separate flammable liquid storage room or warehouse.

Bulk Storage

Bulk storage of flammable liquids requires that a number of safety measures be taken. Essentially, bulk storage of flammable liquids is maintained in storage vessels located either Aboveground or underground. This must be done in accordance with local and national codes and standards.

Underground Storage Tanks

Underground storage tanks require that certain environmental requirements be met so that leaking vessels are quickly identified. In addition, vent pipes of tanks used for liquids with a flash point less than 100°F (37.80°C) should be fitted with flame arrestors. Such an arrestor cools the flame front and the combustion materials to below the ignition temperature of the vapor mixture inside the tank. Flame arrestors, due to their basic construction, are liable to become clogged and, therefore, require periodic maintenance.

Aboveground Storage Tanks

Aboveground storage tanks should be fitted with proper liquid- tight diking, valving, breather venting with flash arrestors, and emergency venting.

Emergency venting is needed whenever liquids are stored in Aboveground tanks, since there is a possibility of a fire exposure to the tank. If this occurs, the rate of liquid volatilization will increase significantly because of the rapid heat input. In turn, this can result in over pressurization and catastrophic failure of the container. This situation can occur even though breather venting is provided, since vapors can evolve much faster than they can be dissipated through the breather vent. Additional emergency venting must be provided to relieve the pressure under this fire exposure situation.

Therefore, it is necessary to provide pressure-relieving valves to avoid catastrophic tank failure. The tank can be designed to fail along the shell-to-roof seam, or devices of a commercial design are available, but even a manhole cover with long bolts will allow the cover to lift if the tank become pressurized.

Breather vents on tanks located inside buildings should extend to the outside so that vapors are not forced into the area around the tank during tank filling. The breather vent line should be fitted with a flame arrestor if the tank contains a liquid with a relatively low flash point. The flame arrestor prevent an external ignition source from entering the tank and causing an explosion.

Finally, it is important to consider the type of tank supports provided. A fire exposure to the tank could cause early collapse of steel supports and severe damage to the tank itself. This can be avoided if the tank supports possess the necessary fire resistance themselves; are protected with fire resistive material; or are prevented from becoming excessively heated.

In this case, installing automatic sprinkler protection beneath the tank is recommended.

Flammable Gases

Ventilation

One of the means of minimizing the potential for a fire or explosion involving flammable gases is to remove the fuel. This is generally accomplished by providing sufficient ventilation to ensure that the fuel-air mixture is far too lean to burn.

The ventilation pick-up point must be located by determining the weight of the gas compared to air. Light gases (specific gravity of less than 1.0) should be exhausted at higher elevations, and heave gases (specific gravity of greater than 1.0) should be exhausted at floor level.

An extension of this same approach is utilized when gases may be present in a container or piping which must be cut or welded. After ventilating the container, an inert gas is usually introduced as an additional safety factor.

Ignition sources such as electrical equipment and room heating in fuel gas cylinder storage areas should preferably be eliminated completely.

Safe Gas Cylinder Storage and Handling

General guidance on the safe storage and handling of compressed gas cylinders are as follows:

1. Protect cylinders against mechanical damage. Provide racks or other means to hold them securely in place and keep valve caps in place except when cylinders are in use.
2. Use specially designed holders for moving cylinders. Do not handle them with electromagnets, ropes, or slings. For manual handling, provide sturdy hand trucks mounting one or two cylinders.
3. Permit only authorized, trained personnel to have access to cylinders. Keep the cylinder storage areas locked to prevent tampering.
4. Locate cylinders where they will not be exposed to excessive heat than can cause pressure build-ups and use only indirect building heating such as steam or hot water.
5. In outdoor storage locations, keep the cylinders off the ground. Surround the area with a substantial wire fence and eliminate all other combustible exposures. Locate the storage as far as practical from buildings.
6. Use piping and fittings suitable for the contained gas and stored pressure. Provide appropriate cylinder valves, regulators, and safety relief devices.
7. Operate cylinder valves and regulators in accordance with the manufacturer's instructions. Keep the valve handles in place for quick emergency shutdown.
8. Maintain cylinders and accessories in good condition. Keep cylinders well painted to prevent corrosion.
9. Provide good ventilation for indoor cylinder areas. Natural ventilation may be adequate for lighter-than-air gases.
10. Provide automatic sprinkler protection if the building construction or contents are combustible or if any fire hazard exposes the gas cylinders.
11. Do not accept delivery from the gas supplier unless cylinder are clearly and legibly marked with the name of the gas contained. Connectors having special sizes and threads are recommended to prevent connecting of the improper gases.
12. Install appropriate electrical equipment in cylinder rooms as specified in the National Electrical Code for.
13. Do not locate cylinders in basements or other low spaces or where escaping gas can migrate to such spaces if the gas is heavier than air.
14. If cylinders are manifolded, provide hydraulic flame arrestors on the discharge side of pressure regulators where any oxygen or air mixture may form in the supply line and there is a possible ignition source, or where oxygen or air may back up in the supply line.

Warehousing

The most efficient way available to store large quantities of material in small areas of floor space is by utilizing bulk storage, solid piling, palletized storage, and rack storage.

As the cost of land continues to rise, warehouse are over-extending their existing storage capacity and new structures are being expanded vertically instead of horizontally.

Methods of Storage

Bulk Storage: Unpacked materials such as pallets, powder, granules, etc., are usually stored in piles and are found in silos, bins, tanks or uncontained piles.

Solid Piling: Cartons, boxes, bales, etc., are piled next to, on top of, and in direct contact with one another. Air spaces and flues exist only where there is imperfect contact with each other.

Palletized Storage: Cubic unit loads of 3 to 4 feet are mounted on their own pallet with the total cube being strong enough to support the weight of additional palletized cubics. The maximum practical height of pallet storage is approximately 30 feet. Palletized storage creates a horizontal air space within the pallets. This air space accelerates the growth and spread of the fire and provides an area shielded from overhead sprinkler protection.

Rack Storage: This is a structural steel framework into which unit loads, generally on pallets, are placed. The height of these storage racks are only limited by the vertical reach of the materials handling Machinery which, like the racks themselves, can be designed for great heights. Generally speaking, most racks are less than 30 feet in height. However, some fully automated warehouse utilizing computerized stacker cranes to offset the high cost of labour, have racks extending up to end exceeding 100 feet in height. In many cases, these automated cranes need less aisle space to operate than conventional methods. Therefore, high rack storage facilities may have narrow (4 foot) aisles allowing fire to more easily expose other material in adjacent racks. The steel rack systems may also be an integral part of the building structure and support the wall and roof of the building.

The commodities stored in high rack storage facilities are normally palletized and sit directly on the rack, not on other pallet cubes. This storage creates both horizontal and vertical spaces or flue which greatly add to the fire spread potential.

The nature or characteristics of the stored materials is a major factor in the fire problems associates with high rack storage. From a fire protection standpoint, the storage of commodities must be considered as a whole, including the packaging. Today's packaging contains increasing amounts of highly combustible materials such as foam plastics.

Stored materials are divided into general hazard classifications based upon the fire behavior of the typical items in each classification. NFPA defines these classifications in Standard #231C on sprinkler protection or rack storage.

National Fire Protection Association standards for storage (NFPA 231 and 231C) group general varieties of stored goods as commodity classifications relative to their heat of combustion, rate of heat release, and rate of flame spread. As used in the NFPA standards, the commodity classifications are Classes I, II, III, and IV which are defined as follows:

Class I: Essentially non-combustible materials on combustible pallets, in ordinary corrugated cartons or in ordinary paper wrappers, with or without pallets. Examples of Class I commodities include beverage in glass or metal containers in ordinary corrugated cartons, electrical motors, and dry insecticides.

Class II: Class II commodities in slated wooden crates, solid wooden crates, or multiple thickness paperboard cartons.

Class III: Wood, paper, natural fiber cloth or low fire hazard plastics with or without pallets. Examples of Class III products are shoes, gloves, magazines, newspapers, wood cabinets, and natural fiber upholstered non- plastic furniture.

Class IV: Class I, II, or III products containing an appreciable amount of ordinary plastic materials or plastic packaging materials. Examples are small electrical appliances, typewriters, telephones, furniture or mattresses with plastic covering or padding, and polyurethane building insulation.

The NFPA standards require increase fire protection measures for each increase in commodity class from I through IV.

22 Highly Flammable Liquids and LPG

Petrochemicals and solvents industries have provided us with a comprehensive range of liquid fuels, solvents, liquefied gases, lubricants, and other imperatives of a modern lifestyle without which our civilization would grind to a halt. The price we pay for having this captive energy on tap is, in safety terms, quite enormous, since they tend to catch fire easily, burn enthusiastically, and create spectacular havoc in a short space of time. Recent fires and explosions in refineries and chemical plants are good examples.

The problem with highly flammable liquids and liquefied petroleum gases is that when accidentally turned loose, they vaporize quickly and link off in the form of an explosive cloud in search of a source of ignition. When the vapor and the ignition source meet, the results can range from a nasty incident to a major disaster, depending upon quantity and location.

The Highly Flammable Liquids and Liquefied Petroleum Gases Regulations 1972, (SI 1972 No. 917), spell out the statutory requirements for safe handling of these terrible twins, and every employer should be aware of the law. The following paragraphs give a summary of the Regulations, but the full text of the official publication should be read and understood by all foremen, supervisors, and managers.

The hazards of the HFL and LPG are broadly similar, we will deal with them separately in summarizing the requirements of the Regulations. The salient points are as follows:

Highly Flammable Liquids (HFL)

1. Interpretation

A highly flammable liquid is defined as any liquid which has a flash point of less than 32°C (92°F), and a fire point of 50°C (122°F). Flash point is the lowest temperature at which sufficient vapor is given off to flash momentarily when a flame is applied. This figure is determined by testing the liquid in a special apparatus designed for the purpose. "Flash point" must not be confused with "fire point", which can be defined as the lowest temperature at which the heat of combustion will generate sufficient quantities of vapor to enable combustion to continue.

2. Storage

All highly flammable liquids must be stored:

a. in suitable fixed storage tanks, in safe positions; or
b. in suitable closed vessels, kept in a store room which is either in a safe position or is a fire-resisting structure; or
c. in suitable closed vessels, kept in a safe position in the open air and protected against direct sunlight; or
d. in the case of a workroom, where the aggregate quantity is less than 50 liters, in suitable closed vessels kept in a suitably placed fire-resisting cupboard or bin.

These regulations do not apply to HFL in the fuel tank of a vehicle, nor to any suitable small, closed container holding less than 500 milliliters, nor to HFL being transported on site or in pipelines, process tanks, etc.

Except where it becomes necessary to gain access, all openings in cupboards, bins or vessels used for storage must be kept closed. Suitable provisions must be made to contain, or to immediately drain off, any highly flammable liquid which may have leaded or spilled.

3. Marking

Every storeroom, bin, cupboard, tank or vessel used for storing HFL must be clearly and boldly labeled "Highly Flammable". Where this is impracticable, then the legend "Highly Flammable Liquid" must be conspicuously displayed as close to the place as possible. This regulation does not apply to spirits for human consumption, (gin, whiskey, etc.) nor to HFL in the fuel tanks of vehicles, closed vessels containing less than 500 milliliters, nor to any aerosol dispenser in which the amount of HFL is less than 45% by weight of the total contents, or 250 g in weight.

4. Spills and Leaks

When possible, HFL must be conveyed within a factory, it should be done by means of a totally enclosed system of pipes and pumps. Where this is not reasonable, they must be conveyed in vessels designed to avoid spills and leaks.

The quantity of HFL in use at any one time must be as small as is possible to reduce the chance for a fire or explosion.

Always we have to make sure that store vessels do not leak, kept closed whenever possible, and are designed to avoid the risk of spillage. Should any spillage occur, it must be contained or drained off safely or absorbed in a safe manner.

Any containers used for holding HFL except for fixed tanks and vessels, must be treated as follows when emptied:

a. refilled for further use, or
b. made free from vapor, or
c. taken to a safe place in the open air, or
d. placed in a fire-resisting structure, cupboard or bin.

All such vessels which have not been made free from vapor must be kept in the storage and kept closed until required for use again.

5. Sources of Ignition

Where there is a likelihood of a dangerous concentration of vapor being present in any work area, there must not be sources of ignition present in that area. Any cotton waste or other material which may be liable to spontaneous ignition, or which is contaminated by HFL, must be placed without delay in a metal container which has a suitable cover, or be removed without delay to a safe place.

Where cellulose nitrate is present in any HFL, all precautions must be taken to prevent the deposit of any solid residue from this source on any surface which is possible to attain a temperature of 120°C.

6. Escape of Vapors

Where possible, steps must be taken to prevent the escape of HFL vapors into any work place. Where this is not possible to accomplish by use of ventilated cabinets or other enclosures, then the work place shall be provided with mechanical exhaust ventilation, to prevent the build-up of dangerous concentrations of vapor.

7. Explosion Pressure Relief

Provision must be made, in a fire-resisting structure, for pressure relief in the event of explosion, or for adequate natural ventilation in a storeroom.

Where pressure-relief provisions are made, the venting of pressure must be directed to a safe place.

8. Means of Escape

Where HFL is stored or used in any room, there must be adequate and safe means of egress for all persons working there.

9. Solid Residues

Where the use of HFL may involve the deposit of solid residues which might be a fire risk:

a. Reasonable steps must be taken to prevent the occurrence of such residues.
b. Where they occur, effective steps must be taken to remove them to a safe place as often as is necessary to prevent danger.

Removal of residues containing cellulose nitrate must not be effected by the use of iron or steel implements.

10. Smoking

Where there is a risk of fire from the presence of HFL, smoking must be absolutely prohibited. Where this applies:

a. Display a clear and conspicuous notice to this effect, as close to the danger as possible.
b. Display a clear and conspicuous notice at every entrance to the site, which prohibits smoking other than in designated areas.

11. Control of Ignition

No HFL shall be ignited other than in plant or apparatus which is suitable for the purpose, and by the proper use of such plant or apparatus. The exceptions to this requirement are:

a. Where the sole purpose is the disposal of waste, in which case it must be burned in a suitable plan or apparatus, by competent personnel and in a safe manner.
b. The firefighting training must be done in a safe manner and under the constant supervision of competent supervisors.

12. Powers of an Inspector

A Health and Safety Executive Inspector may take samples of any material which he considers may be a highly flammable liquid, and submit the samples for testing.

13. Fire Fighting

Where any HFL is manufactured, appropriate means of fighting a fire must be provided use at all times.

Liquefied Petroleum Gas (LPG)

1. Storage and Marking

All LPG which is not in use must be stored as follows:

a. In fixed tanks, in the open air or below ground.
b. In suitable movable tanks or vessels, kept in safe positions in open air.
c. In pipelines and pumps which form part of a totally enclosed pipeline system.
d. In suitable cylinders kept in safe positions in open air, in a storeroom which is ventilated and is of fire-resisting structure.

The amount of LPG present in any work place at any time must be as small as is possible. Cylinders must be properly stored until needed for use and, when emptied, they must be either refilled or returned to the place of storage without delay.

Every tank, vessel, cylinder and storeroom used for storing LPG shall be clearly and boldly marked "High Flammable - LPG" and must be displayed near the area of storage.

There are some exception to these rules. Such as LPG in the fuel tank of any vehicle or engine, quantities less than 500 milliliters in a closed container, and small aerosols.

2. Duties of Employees

Every employee must comply with these rules and must cooperate with the employer to carry them out. Employees also must report immediately any discovered defect in plant or equipments.

Liquefied Petroleum Gas

Liquefied Petroleum Gas has one or two interesting characteristics which should be appreciated because they add to its hazardous nature. LPG is transported in liquid form, the density of the liquid being about half that of water. Once released, the liquid vaporizes rapidly in air and expands to 250 times its volume.

Thus, a unit volume of liquid will quickly revert to 250 volumes of gas, to form a vapor cloud. The density of the gas is greater than that of air, butane is about twice as heavy as air, and propane about 1½ times, so that vapor tends to accumulate in convenient sumps, trenches, drains, gullies, basements, and becomes difficult to remove.

The flammable limits of the vapor are between 2% and 10%, when mixed with air. If below 2% and above 10%, the mixture is either too weak or too rich to ignite. However, when a liquid leakage has occurred, and given rise to a vapor cloud when has expanded to 250 times the volume of the leaked liquid, the outer edge of the cloud will certainly cone within the range of flammability very quickly. When the mixture encounters even a source of ignition, the potential for disaster becomes a reality. A leakage of only one gallon of butane will produce a cloud of vapor at its most explosive concentration (5%) of about 640 cubic feet in volume.

A fire associated with a leakage of LPG should not be extinguished until the leak has been stopped. If the fire is extinguished under other circumstances, the large volume of flammable vapor would form, and the unpredictability of the movement of a gas cloud, could result in an extremely hazardous situation. The turning loose of a large vapor cloud has resulted in many major explosions downwind, and these usually flashback to the original source.

In any escape of LPG, the gas should not normally be ignited or reignited, but there have been cases where ignition or re-ignition have proved desirable, to prevent the formation of a gas cloud. It may be that a sudden escape of LPG could be intentionally ignited within moments of its occurring, and so avoid the formation of a dangerous volume of flammable mixture, but any such action would need to be taken at an extremely early stage and on the decision of the authorities having jurisdiction. At any later stage, the gas should not be ignited and, therefore, such a course is unlikely to arise in the presence of the fire brigade because a considerable gas cloud may have formed before their arrival.

In any escape of LPG which has not ignited, emergency operations must be directed to shutting off the leak at source and, in the meantime, taking all possible steps to prevent ignition and to disperse the gas safely.

If flame impingement is taking place on the vessel, it is important to bear in mind that time is not on the side of the fire service. Experience indicates that after about ten minutes of such impingement, a devastating explosion is a real possibility. Personnel should not be allowed to stand at the front or rear of the vessel. Experience has shown that severe explosion can cause a tank to be propelled like a rocket for a considerable distance.

In the case of horizontal tanks, these may be positioned so that the tank- ends, which are more susceptible to failure under excessive pressure, do not face towards buildings or plants.

Storage of Highly Flammable Liquids

Highly Flammable Liquids are equally problematical in that there are so many of them and they are so widely used in industry.

The HFL and LPG Regulations refers to storage of HFL, and specifies a storeroom of fire-resisting structure, suitable closed vessels, cupboards, or bins.

The purpose of fire resistance in the context of storage is to contain a fire within the structure. During its early stages, it will allow time for people to escape from the surrounding area but it will not protect the contents from a fire outside the structure.

Storerooms can be expected to contain considerable quantities of HFL, and since they are usually unattended, there could be a delay between the start of a fire and its discovery. Thirty minute fire- resistance is therefore specified to protect means of escape.

Workrooms which are required to be of fire-resistance structure must have internal walls which satisfy the thirty minute fire-resistance.

Cabinets and other enclosures are also used for the purpose of containing a fire for as long as possible before affecting the surrounding work area.

Cupboards and bins used to store HFL in workrooms, in quantities not exceeding 50 liters, must meet these standards.

The important features of a suitable cupboard or bin are:

a. Its capacity should not be more than 50 liters in terms of the containers stored within it.
b. Close-fitting doors, or a close-fitting lid must be provided.
c. There should be any ventilation holes or louvres.
d. The internal walls should be of non-combustible materials.
e. Their walls, doors, floor, etc. should be securely bonded to prevent the passage of flame;
f. Their structural members, supports, fastenings, fittings, etc. should be fire-resistant.
g. They must be strong enough to withstand reasonable physical impact, so that its integrity is maintained.
h. Their floor and shelves must be lipped, to contain spillage up to 110% of the volume of the largest vessel normally stored.
i. They must be clearly and boldly marked with the appropriate words, "Highly Flammable Liquid".
j. They should never be placed close to any recognized escape route from the work place.

Storage cupboards for flammable liquids should be properly evaluated before purchase. Those made of plaint sheet of steel offer limited insulation from interior heat, and in the event of a fire inside them, the metal can quickly distort and cause leakage.

Some cabinets have fire-resisting lining, inside the steel walls and doors. These models have a much longer fire-resistance because of more efficient heat insulation.

Containers for storage, dispensing, etc., come in a number of size and guides, and the suppliers publish first-rate catalogs with excellent illustrations. The best of these safety containers are made specifically for the purpose of carrying, dispensing and storing HFL, are properly painted and marked, have carrying handles and a spring-loaded closure system for the filling/pouring nozzle. Some have captive leak proof caps which cannot be lost, plus detachable pouring spouts.

The market should be well explored before a choice is made. The Factories Inspector or the Fire Prevention Office can advise on the required standard which apply to all the equipment mentioned in this chapter. Their advice can save time and money.

Responsibilities of Management

It is obvious from a careful study of the rules and regulations, that those who carry the burden of factory management have a formidable task in handling HFL and LPG. The same care must be taken in other work places. Since the many products which come within the scope of these Regulations can be just a lethal, where they are stored or used. It pays to scan the work place with a fresh eye periodically to check that all the precautions have been taken and people are made aware of what they are handling.

There are plenty of other flammable and easily ignitable chemicals and liquids which can cause havoc, in addition to those with a flash point of less than 32°C. Oils, light greases, paints, solvents, cylinders of flammable gases, aerosols, plastics, polish, wax, and oxidizing agents are just some of them.

Our work places are liberally burdened with cupboards with tins and bottles containing fire hazard materials which catch fire easily and burn with enthusiasm.

Fire Properties of Some Flammable Liquids and Gases

Name	Flash-point (°C)	Flammable Limits	
		Lower	Upper
Flammable Gases			
Acetone	18	2.6%	12.8%
Amyl acetate	25	1.1	7.5
Benzene	-11	1.4	7.1
Carbon disulphide	-30	1.3	50.0
Cyclohexane	-20	1.3	8.0
Ethyl alcohol	13	3.3	19.0
Ethyl acetate	-4	2.2	11.5
Hexane	-22	1.2	7.5
Kerosene	37	0.7	5.0
Methyl alcohol	11	6.7	36.0
Naphtha	-2	0.9	6.0
Pentane	-49	1.5	7.8
Petroleum spirit	-43	1.3	6.0
Toluene	4	1.3	7.0
Acetylene	—	2.6	100.0

Ammonia	—	15.0	28.0
Butane	—	1.8	9.0
Carbon monoxide	—	12.5	74.0
Ethylene	—	2.7	36.0
Hydrogen	—	4.0	75.0
Methane	—	5.0	15.0
Propane	—	2.2	10.0

Note: Flammable limits expressed in percentage by volume in air.

23 Fire Protection in Industrial and Commercial Buildings

Important Aspects of Fire Protection

The most important aspect of fire protection is to prevent the fire from starting. There are only three methods of preventing a fire:

1. Eliminate the fuel.
2. Eliminate the heat.
3. Reduce the level of oxygen to below what it is necessary to support combustion

Control of Oxidizing Agent

An oxygen concentration of approximately 15% is required to sustain combustion, while glowing combustion can occur at much lower oxygen concentrations. Since 20% of air consists of oxygen, it is necessary to significantly dilute air to prevent a fire from starting or continue to burn. Since it is not feasible to lower oxygen concentrations, fire prevention primarily focuses on controlling fuel and heat.

Control of Fuel

A wide variety of fuel is available at many plants. The fuel can be flammable and combustible liquids, flammable gases, paper, packing materials, accumulations of trash, cable insulation, combustible structural materials, interior finishes, and other flammable and combustible materials.

Fire protection activities with respect to fuel should focus on the elimination of unnecessary fuel and attempt to substitute less flammable and combustible materials for those already in use.

The important of good housekeeping should be frequently emphasized to all members of the plant and it is especially important to separate combustibles from heated surfaces such as steam lines, boiler faces, and stationary engine surfaces.

Control of Flammable Liquids

Control of flammable liquids is critical since they are highly combustible. The best way of controlling the fire hazards associated with flammable liquids is to eliminate the liquid itself.

This can sometimes be done if the need for the materials is carefully analyzed and alternatives are considered. For example, a detergent solution might be a suitable substitute for a flammable cleaning solvent or a water based paint may be substituted for oil based paint. Liquids should be selected, whenever possible, that can be stored and used at temperatures that are well below their flash point.

All flammable liquids produce heavier than air vapors which tend to settle at the floor level. These vapors may flow along the floor or ground for long distances and be ignited at some remote point. Therefore, adequate floor level ventilation should be provided near the vapor source. The ventilation may be natural or mechanical. The mechanical ventilation should be used within plants to ensure that the airflow is sufficient under all conditions.

Spills of flammable liquids must be prevented whenever possible. Also it must be controlled quickly so that the amount of liquid involved is minimum. In order to accomplish this, self closing type valves should installed on liquid dispensing lines. Furthermore, to prevent overflow of bulk storage tanks, overflow preventor devices must be installed on the liquid lines. These include float valves, return lines, and specially designed pumps. Approved safety cans should be used for snail quantities of flammable liquids as required by the authorities having jurisdiction.

One practical manner to handle flammable liquids more safely is to separate their area of storage from others. This can be achieved by providing an approved flammable liquid storage cabinet for small quantities of liquids, or by providing a separate flammable liquid storage room or warehouse.

The fire walls provided to separate flammable liquids from other areas must be carefully constructed. The enclosure must be constructed of material which will not allow the liquid to escape from it. The liquid volume must be calculated so that liquid-tight walls and a sill at any door can be provided of sufficient height to contain the liquid.

Aboveground storage tanks should be fitted with proper liquid-tight diking, valving, breather venting with flash arrestors and emergency venting. In addition, vent pipes of those tanks used for liquids with a flash point less than 10°F (37.8°C) must be fitted with flame arrestors. Such an arrestor cools the flame front and the combustion products to below the ignition temperature of the vapor mixture inside the tank. One of the problem with flame arrestors is that they are liable to become clogged and require periodic maintenance.

Specifically designed emergency venting must be provided to relieve the pressure developed in an aboveground tank under a fire exposure to avoid tank failure. The tank can be designed to fail along the shell to roof seam, but even a manhole cover with long bolts will allow the cover to lift if the tank becomes pressured.

Breather vents provided on tanks located inside buildings should extend to the outside so that vapors are not forced into the area where the tank is located during filling operations. If a tank contains a liquid with a low flash point, the breather vent line should be fitted with a flame arrestor to prevent an external ignition source from entering the tank and causing an explosion.

Also it is important to consider the type of tank supports. A fire exposure to the tank could cause early collapse of steel supports and severely damage the tank. This can be avoided if the tank supports are protected with fire-resistive material, or are prevented from getting too hot by installing automatic sprinklers beneath the tank.

Flammable Gases

The major hazard associated with flammable gases confined in a container is explosion. This can happen when container is expanded because of heat. Pressure relief devices are required to reduce the probability of this occurring.

When a container come in contact with flames from an exposing fire, the internal pressure will increase and this causes loss of strength of the steel from which the container is fabricated and ultimately causes container failure. This container failure can create a serious danger to personnel in the form of an explosion even if the contents are not flammable. If the contents are flammable, a very serious fire can also result from the rapid release of the flammable gas.

Once the flammable gases are released from their storage container, they present the hazards of both fire and explosion. Failure to distinguish between the circumstances surrounding these two hazards can result in misapplication of protective measures.

The explosion occurs in the following steps:

1. A flammable gas is released from its container and the gas mixes with the air.
2. When ignited, the flammable mixture burns very fast and produces a great deal of heat.
3. If the air which is heated by the fire is not allowed to expand, the result is a rise in pressure in the enclosure which can cause an explosion.

Since this kind of explosions are caused by combustion, they are called "Combustion explosion".

The primary means of minimizing the potential for a fire or combustion explosion involving flammable gases is to remove the fuel. This is generally accomplished by providing sufficient ventilation to ensure that the fuel-air mixture is too lean to burn, preferably less than 25% of the Lower Explosive Limit (LEL). The ventilation pick-up point must be located by determining the weight of the specific gas as compared to air. Light gases with specific gravity of less than 1.0 should be exhausted near the ceiling, heavy gases with specific gravity of greater than 1.0 should be exhausted at floor level.

Large boilers using gas as the fuel go through a purge cycle to remove any flammable gases that might have been leaking into the firebox. This is done before introducing the pilot flame or additional fuel. If this precaution is not taken, an explosion could occur within the boiler. Same technique is used when gases may be present in a container or piping which must be cut or welded. After ventilating the container, it is a good idea that an inert gas be introduced as an added safety so that ignition cannot occur.

General guidance on the safe storage and handling of compressed gas cylinders include the following:

1. Protect cylinder against any damage. Provide racks or other means to hold them securely in place. Keep valve caps on except when cylinders are in use.

2. Use specially designed holders for moving cylinders. Do not handle them with electromagnets, ropes, or slings. For manual handling, use sturdy hand trucks mounting one or two cylinders at a time.
3. Permit only authorized, trained personnel to have access to the cylinders.
4. Locate cylinders where they will not be exposed to excessive heat. Use only indirect heating system such as steam or hot water.
5. For outdoor storage, keep the cylinders off the ground. Surround the area with a fence and eliminate all other combustible exposures. Locate the storage as far as possible from other structures.
6. Use piping and fitting suitable for the contained gas at stored pressure. Provide adequate valves, regulators, and safety relief devices.
7. Operate cylinder valves and regulators in accordance with the manufacturer's instructions. Keep the valve handles in place for emergency shutdown.
8. Maintain cylinders and accessories in good condition. Keep them well painted to prevent corrosion.
9. Provide adequate ventilation for indoor cylinder areas.
10. Provide automatic sprinkler protection if the building construction or occupancy is combustible, or if the gas cylinders are exposed to any fire hazards.
11. Make sure that cylinder are clearly and legibly marked with the name of the gas contained.
12. Install appropriate electrical equipment in cylinder rooms as recommended buy the authorities having jurisdictions.
13. If the gas is heavier than air, do not locate cylinders in basements or other low spaces.
14. If cylinders are manifolded, provided hydraulic flame arrestors on the discharge side of fuel-gas-pressure regulators to prevent oxygen or air mixture in the supply line.

Control of Ignition Sources

It is essential that any source of ignition such as static electricity, lighting, open flames, cutting, grinding, and welding be prohibited in area where combustible dusts, flammable vapors or gases may be present. Smoking must also be prohibited in hazardous areas. Also the use of the so-called "non-sparking" tools may help further reduce the risk of ignition.

Electrical Equipment

A high percentage of fire is often started by electrical equipment. Since electrical equipment is generally needed in the areas where combustible dust, flammable liquids or flammable gases are stored or used, it is very important that appropriate equipment be selected since ignition can occur easily. Specialized electrical equipment has been developed for three basic classes of hazardous areas:

1. Class I: Areas in which flammable gases or vapors from flammable liquids may be present.
2. Class II: Areas in which combustible dust may be present.

3. Class III: Areas in which ignitable fibers or flying particles are present in quantities sufficient to produce ignitable mixtures.

Class I and Class II areas are each subdivided, depending upon whether or not the hazardous conditions are expected under normal operating conditions or unusual conditions, such as leakage from containers or piping.

If a flammable gas or vapor is present, the area would be Class I. Similarly, if combustible dust is present the area is a Class II. The responsibility of determining the extent of hazardous area rests with the authority having jurisdictions.

The hazards of an area are not the same for all types of flammable gases and vapors, or for all types of combustible dusts. As a result, Class I, II, and III are divided into sub-groups. These groups are:

Class I – Vapors and Gases.

- Group A - Areas containing acetylene.
- Group B - Areas containing hydrogen.
- Group C - Areas containing ethyl ether vapors, ethylene or cyclopropane.
- Group D - Areas containing gasoline, petroleum naphtha, benzene, butane, propane, alcohol, acetone, lacquer solvent vapors or other natural gas.

Class II – Dusts.

- Group A - Areas containing metal dusts, including aluminum, magnesium or other commercial alloys.
- Group B - Areas containing carbon black, coal or coke dust.
- Group C - Areas containing flour, starch or grain dust.

Class III – Ignitable fibers or flyings.

The construction of electrical equipment for Class I areas, explosion-proof equipment, does not prevent penetration of the gases or vapors to the inside of the electrical device.

The term "vapor-proof" is misleading, tests have shown that it is virtually impossible to prevent entrance of gases or vapors, even during minor, normal changes in ambient temperatures. Instead, equipment for Class I areas is required to be constructed so that gas or vapor explosions happening within the apparatus will not ignite surrounding gas or vapor-air mixtures. In order to accomplish this, the enclosures of the equipments are required to withstand vapor-air explosions without rupturing. The Class I groups A-D previously identified are categorized according to ignition temperatures and the explosive pressures developed for the specific gas or vapor. The equipment joints are required to be of the metal-to-metal type with close tolerances to arrest the propagation of flame from the interior of the enclosure to the surrounding areas. In addition, the equipments are required to operate under full load conditions without developing exterior temperatures above the ignition temperature of the flammable gas or vapor to which they are exposed.

Since the equipment design requires close surface tolerances, proper maintenance is essential. Loose equipment covers, missing bolts, or scarred metal surfaces could allow ignition occurring within the equipment to propagate to the general area. Oversized or improper light bulbs, makeshift shields or reflectors on lighting fixtures warrant attention because they can increase the surface temperatures and defeat the equipment design.

Electrical equipments designed for use in the class II areas are to satisfy entirely different objectives. These equipments are designed to prevent dust from entering the equipment enclosures. In addition, the equipment must operate below the ignition temperature of the dust.

Static Electricity

Another potential electrical ignition source for flammable vapors is static electricity. Static electricity is generated when two materials in contact are separated from one another, each becoming oppositely charged. As the frequency of contact and separation increases, so does the static electricity rate. Flammable liquids are poor electrical conductors so that large static charges can be collected and discharged in the form of a static spark.

The amount of static generated can be significantly reduced by minimizing the amount of contact and separation occurring in the flammable liquid. This is accomplished by reducing the splashing in dispensing and handling operations. All areas where flammable liquids are transferred or handled should have complete electrical grounding to reduce the accumulation of static charges.

Grounding equipment should consist of point-type grounding clamps, electrically conductive containers, un-insulated cable and a buss system leading to a ground. This equipment should be tested periodically to ensure its proper working condition.

Welding and Cutting

Welding and cutting operations provide a well recognized ignition source. The vast majority of these fires are the result of improper safeguards during welding and cutting. Most of these fires can be prevented through safe and effective administrative controls.

NFPA 51, Standard for the Design and Installation of Oxygen-Fuel Gas Systems for Welding and Cutting along with Occupational Safety and Health Administration (OSHA) provides guidance on the selection of oxygen-fuel gas equipment. This standard covers cylinders, manifolds, piping systems, protective equipment, hose and regulators and related equipment.

NFPA 51B, Standard for Fire Prevention in Use of Cutting and Welding Procedures, also establishes excellent detailed fire prevention measures to be implemented during welding and cutting operations. These requirements have also been included in OSHA. The major emphasis of a program for controlling the hazards of welding and cutting is placed on management control and protection of combustibles.

NFPA 51B and all major insurance carriers require the use of a hot. work permit system to control the welding as an ignition source. The NFPA standard carefully details the need for trained personnel to issue these permits and perform the work.

To issue the hot work permit, the combustibles in the vicinity must first be evaluated by a trained person. This evaluation should identify those situations which will necessitate either:

1. Relocation of the work.
2. Protection of combustibles within 35 feet of the work
3. Removal of combustibles.

A fire watch should be assigned any time potential for fire exists. The fire watch person should be trained to recognize fire hazards and be able to use portable firefighting equipment. The fire watch person should remain at the scene at all times to watch for fires during the work operation and should stay for a minimum of 30 minutes after work has ceased to ensure no smoldering fire exists.

Hot work should be forbidden in areas containing flammable liquids, gases, or large quantities of easily ignited combustibles. It also should be forbidden in areas of metal covered with combustible material; or on metal sandwich panels with combustible insulation. Fixed fire detection and suppression systems should remain in service to the maximum extend possible during any welding operations.

Material Handling

Material handling at a plant is often provided by industrial trucks that can be propelled by electric motors using storage batteries, or by gasoline, LP-Gas, or diesel-powered engines. These can be an additional source of fuel. In addition, they provide a potentially serious ignition source. They can easily ignited fibers, dusts, vapours, or gases. Unless these vehicles are properly operated and maintained, they can be extremely hazardous. Specially designed trucks are available for hazardous areas. NFPA 505, Fire Safety Standard for Powered Industrial Trucks, lists different types of trucks and identifies the hazardous areas where they must be used.

A system of easily recognized markings has been developed that helps to identify the different types of trucks and the areas where they can be used safely. The appropriate markings are affixed to both sides of the truck so that its type can be identified quickly.

Markers of a corresponding shape are posted at the entrances to hazardous areas. When a truck approaches a hazardous area, the acceptability of the truck for that area can be verified easily by comparing the truck and area markings before the truck enters the area.

The major potential fire hazard associated with gasoline, diesel, and LP-Gas powered trucks are fuel leaks that are ignited by the hot engine, hot muffler, ignition system, other electrical equipment, or external ignition sources. Especial care must be taken in the storage and dispensing of fuels, as with any flammable liquid or gas. Comparatively few fires have occurred in battery-powered trucks.

Proper maintenance of trucks is essential to good fire prevention. A system of regularly scheduled maintenance can greatly reduce the danger of hazardous malfunctions.

Maintenance and repairs to trucks, particularly to their fuel and ignition systems, should be done only in specially designated areas.

Repairs should never be attempted within a hazardous area.

24 Fire Protection in Historical Buildings

The challenges of fire protection retrofitting in historical buildings are:

1. To protect life.
2. To protect property.
3. To make sure that the installation of fire protection systems have minimal impact on the historic features of the building.

Fire protection systems must provide building occupants with maximum safety, and protect property to ensure a minimum loss during fire.

To protect the significant historic features, two principles must be considered. First, the installation and operation of fire safety and suppression systems should have minimal impact on them. Second, these systems and devices should provide them with maximum protection.

In addition, the design of fire protection systems for a historical building must be accomplished within the context of the absolute need to protect life from fire and its effects. Within this context, protection and preservation of features can be accomplished by applying the following standards:

1. Creativity, to ensure the consideration of all possible alternatives that would balance the needs to protect life and property with the overall preservation objectives.
2. Flexibility, to apply and adapt fire protection codes to achieve both the safety and historic preservation objectives.
3. Practicality, to resolve conflicts between fire protection needs and historic preservation objectives.

The complexity of retrofitting a historical building with fire protection systems varies with the degree of its fire safety needs and the historic significance of the building. The more significant the historical features, and the more fire safety risks present, the greater the complexity in achieving a compromise.

Since each project has its own unique issues of preservation and fire safety, it is vital that each project team utilizes a design process that will successfully integrate the needs for fire protection with minimum effect on the historical features. This process must be specific to guide each individual team member, yet flexible to accommodate all of the project.

The Project Team

The project team involved in the renovation and retrofitting of fire protection systems in a historical building, typically include building manager, building occupants, architects, historic preservation specialist, fire protection engineers, and the authorities having jurisdiction.

Specialists must be retained to properly research historic preservation and fire protection issues, and then recommend solutions for a given project. Since the requirements for each area are unique and special, it is necessary that consultation and project coordination be done at the earliest possible time.

So that individual project objectives can be shared and a general plan of action can be developed. These consultations must also occur between the specialists and the State Historic Preservation Office.

The Process of Fire Safety Retrofitting

In designing and installing fire protection systems in historical buildings, the following concerns must carefully be addressed.

1. Historical preservation assessment.
2. Fire protection assessment.
3. Evaluation of objectives.
4. Selection of proposed solutions.
5. Review of proposed solutions.
6. Installation.

The Historical Preservation Assessment:

The Historical Preservation Assessment consists of two parts. A Historic Building Preservation Plan and a Historic Structures Report. These two reports form the frame of reference for the assessment.

The historic preservation specialists examine the building and its site, the zones within the building, and the individual historical features in the building. The building and its site are evaluated based on type, style, use, age, condition, modifications, site context, and historical associations to determine the building's historical significance and historical character. The building zones are evaluated to determine their significance based on their use, original design, public access, integrity, detailing, and materials. Finally, specific features located in the building are evaluated to determine their significance based on uniqueness, materials, detailing, and condition. The result of this evaluation is a determination of the level of significance of the building, zones, and features. Items that are determined to be highly significant require creative and flexible approaches to fire protection retrofitting to preserve their significance.

The Fire Protection Assessment

The purpose of the fire protection assessment is to determine how the building presently performs in the event of a fire, and to define what deficiencies need to be corrected to ensure adequate fire protection for the building and preservation of historical features. It is very important to understand that building codes and life safety codes are guidelines for the fire protection retrofitting of historical buildings. Strict application of the codes may result in the damaging or even total destruction of highly significant features. Therefore, it must be avoided through creative and flexible application of codes.

An alternative approach to codes and standards is an evaluation of buildings performance during a fire. This entails conducting a thorough inspection of the entire property, including stairs, doors, corridors, construction materials, fire hazards, existing fire protection equipment, operational support systems, and the occupancy as part of the overall fire safety.

The result is a logical and reliable determination as to the fire protection needs for the building, occupants, and historical features.

Evaluation of Objectives

As soon as the historical preservation assessment and the fire protection assessment have been completed. The most significant historic features and the most important fire protection objectives are identified and prioritized. Then each priority is analyzed and design solutions explored to meet the objectives of both fire protection and historical preservation. The objective of this analysis is that each team recognizes and understands the values and requirements of each other's program. Such cooperation leads to the resolution of issues of conflicts and paves the road to design the highest quality project, while satisfying both historic preservation and fire protection objectives.

Selection of Proposed Solutions

In the selection phase of the project, the methods of protecting the building occupants and the historical features are decided. Fire protection systems selection, is finalized as the system requirements are balanced against the need for preserving the historical features. Actual fire protection equipments are matched to preservation and restoration techniques and locations, and details of installation are worked out, along with the logistics of accomplishing these activities.

Review of Proposed Solutions

As soon as the solutions from the selection phase have been completed, the project team should review the implications with the authorities having jurisdiction.

This is the point at which the Historic Preservation Officer completed the review requirements administered by the Advisory Council on Historic Preservation.

Once jurisdictional review requirements have been satisfied, the project manager may complete the documentation and issue the drawings for installation.

Installation

During installation, the project team must inspect the work regularly to verify that the original intentions of the design are being met. If construction is required, it is important that photographs be taken prior to commencement of work and be incorporated into the design document to help clarify or pinpoint specific areas warranting special attention. After project is completed, photographs should be taken to document final outcome of the project and also for comparison between pre and post construction.

Code Issues and Fire Safety Systems

The Purpose of Codes in Contemporary Buildings

Modern construction is required to adhere to a number of codes and standards. Specific codes are written for building construction, mechanical, plumbing, electrical, and fire protection systems.

The typical contemporary building is designed for life safety and fire protection as well as structural, mechanical, and electrical systems requiring materials and methods of construction that are well known to the architects and engineers.

Building codes and fire protection codes establish minimum standards for building construction. Most codes determine allowable construction techniques or materials by weighing the degree of safety provided by the structure against the degree of hazard presented by the occupants and by taking into account such factors as installed fire protection systems.

Building Codes

The purpose of building codes is to provide minimum standards to safeguard life, property, and public welfare by regulating and controlling the design, construction, occupancy, and maintenance of all buildings and structure within its jurisdiction. The building codes define specific criteria for buildings based on occupancy, height, location, and type of construction. In addition, these codes specify detailed regulations such as existing fire extinguishing systems, wall and ceiling covering, elevators and others.

There are three building codes that are enforced throughout the country:

1. The Basic Building Code, used mostly in the northeast.
2. The Southern Building Code, used mostly in the southeast.
3. The Uniform Building Code, used mostly in the west.

In addition, some of the larger cities such as New York, San Francisco, and others have adopted their own set of codes. Most codes also include appendices that reference industry standards prepared by manufacturers or research associations such as American Society for Testing and Materials (ASTM), American National Standards Institute (ANSI), Under writers Laboratories (UL), and Factory Mutual (FM), American Society of Mechanical Engineers (ASME), National Fire Protection Association (NFPA), and others.

Fire Protection Codes

The most widely recognized Fire Protection Codes are published by the National Fire Protection Association (NFPA). NFPA focuses on life safety as well as fire suppression systems to protect the property.

The Life Safety Code addresses those items in building design and operation which affect safe egress from a building. Means of egress, building fire safety features, and fire suppression systems are just a few of areas which fall under life safety codes.

The specific requirements for each of these areas varies depending on the occupancy classification of the building. Occupancy is defined as the purpose for which a building is used. Therefore, the intended reuse or continued use of the historic building will determine the design and engineering of the life safety solutions.

Means of egress is defined as a continuous and unobstructed way of travel from any point in a building to a public exit. The means of egress is an integral component of life safety systems and it makes safe evacuation possible.

Building Safety Features are incorporated into the structure. The code addresses specific requirements for installation of vertical and horizontal fire-rated partitions, smoke barriers, and interior finish materials.

Building fire protection systems include fire alarm, detection and communication devices, automatic sprinkler systems, and other extinguishing equipment.

Codes in Historic Buildings

Compliance with building and fire codes is more difficult for historical buildings than for new construction. In fact, there are no specific codes for installation of fire protection systems in historical buildings. The building codes make special provisions allowing the authority having jurisdiction to recommend code requirements for construction, alterations, and the repair of historical properties.

NFPA has published recommended codes for fire protection in historical buildings. The purpose of these practices is to provide guidance, requirements, and information regarding the identification of fire hazards, and recommendations for planning, design approaches, and solutions appropriate for the historical building relative to fire protection and fire prevention needs.

Alternative Approaches to the Codes

Qualitative risk assessment is a preferred method for determination of proper fire protection retrofitting in historical buildings. Risk assessment incorporates the identification and evaluation of the conditions of the building, the potential consequences, and the associated risk to the occupants, property, and the mission. All positive and negative features must be taken into consideration in the building evaluation. This approach will project the expected performance of the property during a fire from which an assessment of the risk to life and property can be formulated. Once a risk assessment has been determined, the levels of acceptable risk must be agreed upon.

Risk is the potential for harm or negative consequences of an event. The objectives of control of risk from fires is to reduce the probability and consequences of events leading to and resulting from fires to an acceptable level.

This systems approach using risk assessment techniques should be applied to determine how conditions are to be corrected. Such techniques should be used by project managers when developing abatement procedures for conditions. Recommendations are to be based on the exposure to risk of fire loss.

Each historical building must be individually evaluated from a fire protection system standpoint and acceptable levels of risk must be established. The design team should research the existing infrastructure early in the planning stages to determine the value of a fire protection system inherent to the existing condition compared to modern building standards. Then the design team can approach the applicable fire and life safety codes as a guide.

Then the team should design and engineer solutions which apply risk assessment techniques with the need to protect the historical parts of the building. Resolution of potential conflicts can be coordinated through the authorities having jurisdiction and the historic preservation specialists.

The design team should pay special attention to the following components of the building:

Corridors

Corridors that lead to an exit are typically a component of a means of egress. The pathway to an exit must be wide enough to accommodate egress of building occupants and be located in such a manner that travel distances to the exit are the shortest and the most direct. Corridors should be arranged so that if a fire blocks access to an exit in one direction, the occupants could access a second exit in opposite direction.

The historical significance of the building within its corridors must be maintained without sacrificing fire protection requirements. The permanent alteration or removal of the historically significant ceiling, floor, wall, or doors are prohibited so would be installing new ceiling, floor, wall, doors, or openings that would permanently alter the appearance of the historically significant building.

Stairs

Stairs are a primary component of a means of egress and serve as the exit or protected pathway between the corridor and exit discharge. Stairs that serve as exits usually require separation from other spaces by fire-rated enclosures. If a stair in a historical building is found to be historically significant, then the project team must come up with a solution that preserves the stair, yet also provides building occupants a safe means of egress.

Doors

Doors serve as an interface between exit corridors and exits, as well as an interface between exits and exit discharges. In both cases, the doors are typically fire rated and kept in the normally closed position in order to maintain the fire rated integrity of the means of egress. In some situations, it is impossible to keep these doors in the normally closed position due to high frequency travel and in some cases where the doors interfere with the historic integrity of the building. In these instances, electromagnetic door hold-open devices can be used to keep the doors in the normally open position. Upon receipt of a fire alarm signal, the door hold-open devices will automatically de-energize and release the doors so they will close and latch.

An alternate method to provide a fire-rated equivalency for a historic door is to install a fire sprinkler on each side of the door. This method could be utilized in lieu of replacing the door.

Materials of Construction

The materials used in the construction of a modern building are required to comply with various fire and building codes. These materials include many components and systems used for interior walls, and the finishes. However, the materials used in the building of a historical significance may or may not comply with current fire and building standards.

In many instances, historical buildings involve the use of masonry walls with plaster, which are inherently fire resistance. The project team should identify equivalent fire resistive ratings for the various existing materials in the Assessment and Evaluation stages of the fire protection retrofitting process and then select the appropriate means, as required, to create comparable ratings.

Automatic Fire Sprinkler Protection

Where the automatic fire sprinkler protection is required for life safety or protection of property, careful planning is required to ensure that its installation is conducted with minimal disturbance and damage to the historic building.

Piping should be concealed where it is possible. Where piping must run exposed, the least intrusive methods should be planned, including furring of walls and painting piping to match existing ceilings and walls. Although the sprinkler heads themselves must be exposed, there are various methods of installation where their presence can either be minimal or discreetly hidden and not necessarily detract from the aesthetic or historic nature of the property.

Fire Extinguishers

Fire extinguishers must be installed without the permanent alteration of the appearance of the historical building.

Surface mounted fire extinguisher cabinets must be used in areas where recessed cabinets would alter the significant historical fabric, such as marble wainscoting.

Recess mounted fire extinguisher cabinets must be used where possible.

Care must be taken in the selection of fire cabinets style that is least obtrusive to the surrounding historical fabric.

Fire Alarm and Detection Systems

Fire alarm and detection devices, because of their inherent function, are more difficult to adapt to historical properties. These devices usually cannot be hidden, therefore their installation must be closely coordinated between the authorities having jurisdiction and the historic preservation specialist. Conduit installation and locations should be treated similarly to that of sprinkler piping.

Installation of smoke and heat detectors and required electrical conduits must be done in a manner that they are not prominent and do not affect the significant historical fabric of the building.

Conclusion

Fire safety improvements support historic preservation objectives, as such improvements ultimately will protect the property from extensive damage in a fire. In most cases, these improvements can be accomplished without significantly altering the historic features of the building. Construction codes define a set of minimum requirements for the design and construction of a building and should be used only as guidelines. Each historical property must be individually evaluated for a fire safety system and acceptable level of risk must be established. Fire safety cannot rely on a single safeguard, but is based on several components.

Any one component can fail, and so multiple safeguards are necessary to assure a reasonable degree of life safety.

25 Early Detection Alarms

The basic fire protection principles are automatic fire detection and fire suppression systems. Fire alarm systems are designed to alert the occupants and save lives. In most cases lack of alarm systems have caused needless injuries and loss of lives to the occupants. Furthermore, because of building configuration, construction material, and contents the presence of automatic suppression systems does not always contain a rapidly growing and intense fire. In these cases, early warning from smoke detecting systems can save lives and also can reduce property losses by prompting an earlier response.

When fire detection and fire suppression systems are properly designed and installed, they can do both save life and property.

There are three main objectives in fire protection.

1. Protecting life.
2. Protecting property.
3. Protecting the continuity of the business.

Saving life is the primary function of early warning smoke detecting systems. Fire drills, building evacuation schemes, and the presence of life-safety personnel are a major part of the total fire protection system but they cannot be substitute for an early warning smoke detection system which saves lives.

The role of smoke detectors is not significant in fighting fires. Therefore, to protect the building we must employ sprinkler systems to extinguish fires. Their primary function is to contain the blaze until the fire department arrives. Most standard sprinkler heads are not activated until a fire produces at least 16°F heat. In most cases a great deal of damage is done by then.

The early warning given by a smoke detector often summons the building's own staff in time to handle a small fire with minimal loss. A conventional smoke detector usually detects fires long before a sprinkler is activated. The smoke detection systems can be set to alarm virtually at the first sight of smoke. Also special heat detectors are available that use either metal conductors or fiber optics, and flame detectors that combine ultraviolet and infrared sensors and, in some cases, utilize fiber optic networks. Even when viewed solely as life safety devices, early warning smoke detectors can play a major part in the property protection. Since an early warning system can get the people out quickly and safely, the fire department is freed from search and rescue operations and can start immediately suppressing the fire. Sprinklers, smoke detectors, heat sensors, and flame detection devices can be designed to complement each other and reduce the potential for loss.

For fires that produce a great deal of heat almost immediately, fast-response sprinklers respond almost as quickly as heat detectors. However, many fires smolder for minutes or sometimes hours before flaring up. These types of fires can be caught by detectors long before they reach the flaming stage, which is why smoke detectors often are invaluable in loss prevention. Many businesses understand the benefits of smoke detection systems and hire consultants to help them implement such systems on their properties. Unfortunately, some business managers do not go beyond the minimum requirements mandated by local fire and building codes. This is an area in which the insurance industry could be a driving force in urging companies to install heat and smoke detection systems that complements sprinklers for maximum protection of life, property and business continuity. Reduction on insurance premiums can be a powerful incentive for additional protection. When a company plans to install an early-warning detection system, it should conduct a risk analysis to establish the specific hazards and protection needs of that building. The risk analysis should determine the fire potential of the area, the kind of fire, and the type and location of early warning detectors for the best possible loss protection.

The best risk assessment views each building in terms of its occupancy and use. What is the tenant doing on this floor? What vital records are being stored, and how expensive are to replace them? What is the overall importance of the building's use and equipment? Then all of the questions must be evaluated and a detection system be custom designed to do the most good for both the occupant and the building. A case can be made for detectors in nearly every building. In a high rise office building, the usual fire risks are relatively snail, but if the loss potential is high, an early detection system can be installed to catch a fire in the earliest stage.

Different approaches are necessary in a hospital or nursing home, where occupants cannot remove themselves, or in a hotel, where they may be asleep when the alarm goes off. In an industrial situation, minimal protection may be adequate for open areas, but specific processes may warrant special protection due to the equipment's high value or long replacement time. When a new office building is planned, only a very general risk assessment can be performed. Detection systems installed during construction typically have generic specifications and cover only the core of the building. However, not all systems can be easily tailored to meet the special needs of future occupants. Since some tenants will pose higher risks that others, and the arrangement of offices and floors is not known, the best solution is to install a versatile system that can accommodate future changes. A number of highly flexible microprocessor controlled systems are now available. Since intelligence is centered in the control panel rather that the devices themselves, these systems allow for changes to be made in the types of detectors used at different locations in the building. The addition and deletion of detector heads and the re-programming of detector sensitivities is done directly from the control panel. With microprocessor controlled systems, fire protection arrangements can be changed and specific needs can be addressed as they arise. Some insurers hesitate to offer premium rate inducements for smoke detection because they feel the potential for false alarm is too great. As with risk assessments, the nuisance alarm question also involves a consideration of how large a fire can be tolerated before the alarm must be sounded. The higher the flame that can be tolerated, the easier it is to eliminate false alarms. On the other hand, the earlier the detection, the greater chances are the equipment will misread the signals. The relative cost of nuisance alarms must be weighted against the risk of giving a fire a good head start. The fire protection industry is making progress in developing detectors that are sensitive enough to provide meaningful early warning while minimizing unwanted alarms. Among the newer techniques aimed at making detection equipment more discriminating are alarm verification, positive alarm sequencing, variable sensitivity settings and detectors that maintain a constant sensitivity over time.

One of the new technique is algorithm processing that distinguishes the true fire signature. In a high-tech approach to eliminating nuisance alarms, at least one manufacturer is working on an algorithmic approach based on the new generation of addressable, microprocessor controlled detectors. The aim is to develop a smart control panel that can identify specific fire fingerprints or signatures for each detector location. Eventually a menu will be developed, and the panels will function based on the typical scenario assigned for each detector.

In developing such a system, researchers have computer logged enormous amounts of data on possible environmental conditions to create algorithms showing the difference between nuisance situations and a true fire signature. The problem is that so many variables are inherent in each fire situation. The solution may be for the alarm to activate when any one of a combination of conditions is detected, for example, the presence of smoke accompanied by a rise in temperature, however, when design, installation, and maintenance are handled properly, the incidence of false alarms is greatly reduced, even with conventional systems. Investigating application needs and hazards, selecting the right detectors, putting them in the right locations, and properly testing and maintaining the system will significantly reduce or eliminate false alarms. The fire protection industry also is continually improving its products, which are much less likely to signal a false alarm that they were five or ten years ago. The cornerstones of a good system are reliable equipment, proper application of appropriate detectors in each location and regular maintenance to keep the system working as it was when the acceptance tests were run. Expertise in all these areas is important.

A great deal has been done to improve fire protection but a great deal more is waiting to be done. The installation of sprinklers is an excellent step toward better property protection. The addition of interactive detection systems, especially the development of smart systems, could save lives and properties.

But all of this requires extra investment and little gets done without incentives. All too often it takes a disastrous fire, with tremendous property damage and tragic loss of life, to prompt the enhancement of local fire codes. However, the insurance industry could inspire such changes by urging and rewarding the installation of detection systems to complement existing suppression systems. The result would be of enormous benefit to property owners as well as to the insurance industry.

26 Fire Protection for Aircraft Hangars

Aircraft hanger represents a huge loss potential with the primary hazard being the ignition of a fuel spilled by an aircraft parked in the hangar. An aircraft is extremely vulnerable to damage caused by fire and the hangar itself is susceptible to fire damage, because its steel supported roof will buckle and possibly collapse as a result of exposure to high temperatures.

Extinguishing the fire

Fast extinguishing of fire is vital if loss is to be controlled. Even when it is not possible to save an aircraft, it is necessary to control and extinguish the fire to protect the hangar. Since flammable liquid is the principal fire hazard, extinguishing systems must employ foam as the preferred means of protection. While the aircraft and the building structure are cooled by the water, the burning fuel is smothered by the foam. Water has higher density requirements than fuel, resulting in water carrying the fuel to other areas and spreading the fire. It is not practical to mount fixed fire protection directly under the aircraft where there is extreme fire hazard. Therefore, three kind of extinguishing systems are recommended for total coverage all employing foam.

1. Monitor Nozzle System.
   This system is automatically activated by highly sensitive optical detectors and is used to protect the underside of the aircraft.
2. Roof Deluge Sprinkler System.
   This system is automatically activated by heat detectors providing building protection in the event the monitor nozzle system fails to extinguish the fire.
3. Hand Held Hose Lines.
   This system is provided for manual firefighting operations.

Since building protection is provided by a deluge system large amounts of water and foam concentrate are required These must be properly managed by a pumping facilities to meet high volume demands.

Foam Deluge System

The first detection normally is done by the monitor detection system utilizing infra-red or ultra-violet detectors. These systems are capable of detecting a fire in its early stages. However, false trips with such detection systems are a good possibility and must be seriously considered. Manual activation on the monitor system must also be provided.

In a typical system actuation, a signal from the monitor detector is received by the control panel to open both flow control valves supplying water and foam to the monitor nozzles and sound the alarm. Foam concentrate is proportioned into the water and mixed in the supply piping. The Flow Control Valve operates both as a deluge valve and a downstream pressure regulating valve to insure the correct proportion of water and foam concentrate regardless of varying supply pressure. The monitor nozzles move in a programmed path to cover the aircraft, the under-aircraft areas, and areas that might contain a fuel spill. Hose lines supplied by manual valves are located at convenient places for manual firefighting. In the event the monitor nozzles do not control the fire, the temperature will increase at the roof. This condition will cause the flow control valve to open. Unsealed upright sprinklers then distribute the foam over the general area. Operation of the roof system will trip the monitor system in the event that it has not already operated. That operation may also be designed to actuate systems in adjacent areas.

27 Smoke Detectors

A smoke detector can provide vital early warning to protect you and your family from death and injury in case of a fire in your home. Almost 6,000 people in the U.S. die every year as a result of structural fires, making it fourth as a cause of accidental death. Of these fatalities, more than 80% occur in family homes. Fortunately, there are fewer fire deaths today than there were 10 years ago, even though, there are twice as many fires. There is a simple explanation for the declining number of deaths in the face of increasing dollar losses. It is called the smoke detector.

Smoke Detectors

There are two basic types of smoke detectors.

1. An ionization unit contains a small amount, less than one microcuric, of a radioactive substance. The decaying material fills one or two sampling chambers with ions of nitrogen and oxygen, which support a very small electrical current that keeps the alarm silent. Should smoke enter the chamber, the ions attach to the particles, and the current path breaks down, sounding the alarm.
2. Photoelectric smoke alarms rely on light scattering to detect smoke. As long as light emitted periodically by a diode fails to strike a photocell set out of line of the light path, the alarm stays off. But if smoke enters the light-tight chamber, the particles will reflect and refract the light onto the photocell, sounding the alarm.

The ionization detector responds better to smoke consisting of numerous small, less than one micron, particles typical of the emissions from a blazing fire. It will, for example, respond quickly to toxic gases given off by certain burning plastics. Unfortunately, it's also quite sensitive to the fumes produced in a busy kitchen.

A photoelectric detector responds more quickly to the large, greater than one micron, particles produced by smoldering fires. It will warn earlier of the sort of fire caused by a careless smoker or by spontaneous combustion. And it is much less prone to sound nuisance alarms. The main reason for failing is lack of insensitivity to fumes and to black particles, which absorb rather than reflect light.

It seems that the photoelectric detector would probably be the better choice. According to the National Fire Protection Association, nearly all house fires give off dense smoke before bursting into flames or creating toxic fumes. In some instances, the lag time between smoke and a hazardous atmosphere may be as little as one minute, so early warning is vital.

Fortunately, you don't have to choose between the technologies. Several companies produce alarms that use both sampling methods, and the combination is considerably more effective than either alone. Some companies even offer ionization or photoelectric detectors equipped with heat sensors. Except in special circumstances, though, this is probably an unnecessary option. For the home owner, the ionization photoelectric combination gives the earliest warning, since smoke and fumes precede any significant rise in temperature.

New homes are now equipped with hard-wired interconnected smoke detectors as a matter of course, but the 9 VDC battery type is still widely used in existing homes. Either can be effective, but the wired units do have advantages. First, they don't require annual battery replacement. But equally important is their easy interconnection. While they're being wired, the units can all be set to sound when one detector senses smoke. Thus people sleeping in one part of the house will be sure to be aware of a fire in a remote area. It is possible to interconnect battery-type. Alarms, but stringing wire defeats their main advantage; simple installation. Of course, hard-wired detectors are dependent on the house's electrical system, while battery-powered alarms will sound even during a power outage. If you are building in an area prone to power loss, you might consider a hard-wired system with battery backup. The weak-battery warning alarms in these units are designed for a particular battery's characteristics, and substituting a different type may corrupt the system.

Many other features are available to the discerning smoke detector shopper. One particularly valuable option is a light, flashing or continuous that indicates that the detector is operating normally. All smoke alarms have manual test circuits, but a quick glance is much easier that reaching up to push a button. When you're shopping, it's also worth checking the sound level of the alarm. The minimum is 85 decibels, but some detectors sound at as high as 95 decibels, which might be useful in a basement but would be unwelcome in your living room. Other models offer lights to help you find your way out of the building, and even computer-synthesized voices giving exit instructions.

Location of Smoke Detectors

Though every new smoke detector includes detailed placement instruction, an understanding of the principles of fire detection can help you tailor those guidelines to your home.

About 70% of all residential fire fatalities occur between 8:00 PM and 8:00 AM, so the main danger from fire is that sleeping people will be overcome by smoke or toxic fumes. For that reason, smoke detectors should be placed in the path smoke would take to reach the bedrooms from the rest of the house and in locations close enough to the bedrooms to rouse everyone. Thus, depending on your home's layout, you may need from one to several detectors. For example, a single detector located on the ceiling outside the bedroom doors might adequately protect a single storey house with all the bedrooms in one wing. On the other hand, if the master bedroom is isolated from the other bedrooms, there should be and additional detector outside that door. Two-storey houses should have a detector on each floor. One should be installed outside the bedrooms on the second floor, and another should be located near the stairwell on the first floor. If there is a basement, it would also be a good idea to place one near the stairway that leads up to the first floor. It isn't necessary to have a detector at each level of a split level house unless doors separate the areas. All detectors wired to household current and battery-operated detectors, if possible, should be interconnected so they all sound when one senses smoke. Additional detectors—perhaps in the bedrooms if doors are kept closed, or in the living room—would increase the safety margin slightly. Avoid putting detectors in the kitchen, in the garage, or near fireplace, since false alarms are likely to occur in these place. As for the specifics of installation, because smoke rises, detectors usually work best when attached to the ceiling. However, if the ceiling is insulated or radiantly heated, place them on the wall between 4" and 12" from the ceiling has exposed joists or beams, place the detector on the bottom of one. On a ceiling with a slope greater that 1:8, the detector should be set within 4" and 12" vertically and 4" and Laterally of the highest point. Detectors should never be installed with 4" of a wall-ceiling junction or corner, either of which smoke may bypass. Likewise, don't position detectors in their path of heating or air- conditioning vents.

Evacuation Plan

Though installing a smoke detector is the first step toward fire safety in your home, knowing what to do when the alarm goes off is an important. You should have an evacuation plan, and everyone in the household should understand and practice it. Though fire drills may see a little silly at the time, practice is important because of the circumstances in which fire danger is likely to arise. When a smoke alarm awakens a sleeping family in the middle of the night, few are likely to be clearheaded. Groggy, frightened people are safest following a standard routine. Fire drills are particularly important for small children and the elderly, who are more than twice as likely to die in a residential fire as those between 10 and 60. Everyone should understand that the most important action in the event of a fire is to get out of the house as quickly and safely as possible. There is no time to get dressed, collect pets or personal possessions, or even call the fire department.

Plan two exits from every room. For second-storey bedrooms, that may mean installing a ladder at a window. There are commercial rope and chain models. The first choice exit through the house is the normal one, but check it before proceeding. Before opening a door to a hall, look for smoke coming under it, and place a hand against it to check for warmth. If either test is positive, use the window exit instead.

Since smoke rises, practice crawling to avoid it. Because electricity is likely to fail during a fire. You should also learn to find you way out without the aid of lights. It might even be wise to try the route blindfolded. If at all possible, people leaving the house should close doors behind them to slow the fire's progress. Everyone should also know to drop to the floor and roll back and forth to extinguish fire on clothing. Once safely outside, all should gather at a predetermined place so you can assess whether everyone has gotten out. Often people die when they go to a burning building to save someone who is already out.

Ionizing smoke detectors contain a small amount of radioactive material that under certain conditions may be considered a health hazard. Theories on the effects of low level radiation can be very controversial. Some scientists believe that low level radiation may be beneficial and some believe that its effects can be dangerous. With such a lack of consensus, it's probably best to be cautious. Under any condition you would be safer with an ionizing smoke detector that with no smoke detector at all. In fact, in most cases it is good having an ionizing detector as well as a photoelectric model.

The risk from fire is certainly greater than that of radiation from a smoke detector. As a matter of fact, there is almost no risk to a home owner unless the foil packaging around the radioactive isotope is broken.

Even so, you should handle an ionizing detector carefully and dispose of it only by returning it back to its manufacturer. Never take an ionizing detector apart, and don't keep it in the house if it is damaged. Use it with good sense and a lot of care.

Fire Extinguishers

Though most life threatening fires occur when people are sleeping, a fire extinguisher may still save lives and can certainly reduce damage. Like for one with at least five pounds, and preferably 10 pounds, capacity, with an A-B-C rating. Keep it in a central location near the kitchen (for cooking-related problems) and the living room, if there is a fireplace or woodstove. It's also a good idea to keep another in the garage.

Anyone who may use a fire extinguisher should read the instructions carefully and commit them to memory.

Technique is very important in getting the most from an extinguisher, and the approach varies depending on the type of fire.

28 Fire Protection at the Job Site

Construction sites are notorious for their fire hazard. In 1994, in Southern California over 80 apartments, almost finished, were destroyed by sparks from a welding torch. In Michigan a shopping center under construction was destroyed by the crew's open fire to keep warm.

Every manager would like to do his best to ensure that the building will get built with a minimum amount of disruption. The followings are several suggestions that may help to achieve that goal.

Safety at the Site

All contract entering the site must go through a safety orientation program. This program must consist of lectures on site safety, a short video about the site policies and procedures, a pamphlet explaining rules and regulations, and follow it with a short test. This not only reinforces your policies, but helps you document the training and gives you the clout you may need down the road in the event a person flagrantly violates procedures and is dismissed from the site.

There must be firm rules and regulations fire safety on the site for all employees before you can show credibility to the outside contractors.

Hazardous work

Hot Work and Burning are major sources of fire and accidents on a construction site. All hot work should be done through a permit system supervised by the site safety department, fire brigade, project engineers, or project manager. Do not depend on contract workers to precisely follow your hot work procedures. Their operations must be monitored by inspectors. Things to look for are charged and certified fire extinguishers, clearance around the site, and a fire watch whose sole duty is to monitor the safety of the job and have the capability to summon help and use portable fire extinguishing equipment. Also, ensure that compressed gas cylinders are properly secured and standing upright when being used or stored.

Flammable and combustible materials such as fuel for vehicles, epoxies and coatings for floors and walls, glues for joining plastic pipe, and gasses for hot work, are usually found in fairly large quantities on a construction site. Anywhere there is a storage or dispensing of these materials, a properly rated fire extinguisher should be on hand. Limit portable storage cans containing flammable liquids to 5 gallons maximum and use only approved safety cans and also limit inside storage cabinets. Outside, store flammables at least 20 feet from the building. It's good practice to have the workers remove all flammable liquids from the work place at the end of each day so responding emergency services find no surprises during a fire. Of course, ignition sources should be avoided or eliminated by guarding against sparks, smoking materials, and place heaters.

Housekeeping

All scrap and trash should be removed from the building on daily basis. Don't allow trash and debris to block egress from the building. Ensure that oil-soaked rags and other similar types of materials are placed in metal cans with self-closing lids.

Dumpsters should be provided by contractors to handle their trash without having to use "company" trash receptacles. Dumpsters containing combustibles should be kept a minimum of 25 feet from and building, trailer, substation or overhead pipe rack or power line. If a contractor is using a trash chute from the upper floors to the dumpster, the chute should be removed from the dumpster at the end of each day. Since the chute will act as a chimney or flue in case of a fire and causes the fire to spread to other floors in the building. There are many documented cases where a dumpster fire has become a major building fire.

Fire Prevention and Protection

Movement in and out of the construction area as well as the building itself must not be impaired so that workers can evacuate if necessary and to give easy access to incoming emergency services. All buildings must be accessible from all sides for emergency vehicles to gain access. A good rule of thumb is a 20 foot fire aisle with at least 14 feet of overhead clearance.

In buildings under renovation, the existing sprinkler and standpipe systems should remain active until all interior demolition has taken place. If the fire protection system is being replaced, the new system should be completed and activated prior to occupancy of the area. If the old system is to remain, the system should stay in service as reconstruction begins and progresses. However, activation of the sprinkler system should be handled as a priority in all cases.

Fire extinguishers should be strategically placed throughout the site and clearly marked for easy location identification.

The site fire protection equipment such as hydrants, hose houses, wheeled extinguishers, etc. should not be blocked or its use impeded in any way. In the event that a tie-in needs to be made to the fire water main system, a permit should be obtained and the proper shutdown arrangement made. Except in the event of an emergency, use of any site fire protection equipment should be considered a safety violation.

There are a few additional precautions to consider regarding hazardous materials. If there is a need to store liquids, gasses, solvents, and other fire hazards; special permission should be secured from authorities having jurisdictions.

The bulk storage should be at least 50 feet from any building and no more that a working day's supply should be kept in the building. If the materials cause harmful vapours or nuisance odours, adequate ventilation must be provided.

Any modifications to the existing gas piping systems should be done with the supply line shut. Tar kettles for roofing should be kept on the ground and at least 25 feet from the building. At least one fire extinguisher must be provided. Tar kettles must have a self-closing lid and must not be permitted on the roof.

Planning Your Construction Site

Fire prevention preplanning is as imperative with a construction site as with any occupied building. If your first line of defence against fire is the local municipal fire department, invite them in periodically to look around and get familiar with the construction site. If you have a fire brigade or industrial fire department, they must become familiar with the site also.

The following checklist may assist you in making your job site safer and more productive.

1. Twenty-foot emergency access.
2. Egress is clear for building evacuation.
3. No combustibles in unprotected area.
4. All hot work procedures complied with.
5. One 2A extinguisher per 4,000 sq ft.
6. Temporary heating system to be MFD/UL/FM approved.
7. Lay down area must be 25 ft. clear of building.
8. Dumpsters must be 25 ft. clear of building.
9. Good housekeeping practices.
10. Flammable liquids must be 50 ft. from building.
11. Trailers must have fire protection.
12. No parking in the fire lane.
13. No unauthorized fire protection hookups.
14. Adequate ventilation for hazardous materials.

29 Fire Protection for Storage

In today's warehouses, one probably sees a modern storage facility or distribution center with the latest products and materials-handling technology. However, these state-of-the- art facilities loaded with products are protected with a sprinkler system designed 20 or more years ago?

The most effective line of defence against fire in a warehouse is an automatic fire sprinkler system. Yet, despite all of the statistics backing up sprinklers performance, some people resist their installation because of the expense.

From 1989 to 1993, Factory Mutual Engineering group studied sprinkler system protection for about 400 storage fires with a total property loss of more than $391 million. Of these fires, 106 were at facilities with sprinkler system. Total loss was 28 million dollars.

In comparison, there were 148 storage fires at facilities without sprinkler protection. The total property loss was more than $197 million, making the average property loss ratio of unsprinklered facilities to sprinkler facilities five to one.

Automatic sprinkler systems for any warehouse is designed based on product stored and the storage arrangement. Both the product itself and its packaging influence the total amount of combustible material and the rate at which it burns. The storage arrangement and storage height influence the burning rate and the intensity and size of fire. The building height also influences protection requirements.

Flammable liquid once stored in metal cans are now often stored in plastic containers. Burning plastic containers will soften, melt, and release their contents, creating a liquid spill fire.

Packaging materials are changed from paper to a plastic, such as polystyrene foam or foamed-in-place packaging. Plastic packaging significantly increases the combustibility of a product.

Open frame racks are equipped with solid shelves for ease of handling. Solid shelf racks obstruct sprinkler water penetration from the ceiling sprinklers. Design modification may be needed to get water to the seat of the fire.

Choosing the Right Sprinklers

Protection of all commodities up to and including plastics, no special hazards such as flammable liquids or aerosols, and consider each sprinkler option from the standpoint of protection level, flexibility, cost, and whether or not it's a good candidate for a retrofit.

Quick Response Sprinklers

Until the Early Suppression-Fast Response (ESFR) sprinkler was developed in the mid 1980's all sprinklers operated in fire control mode. Sprinklers limited fire spread by prewetting a surrounding commodity until the fire was manually extinguished. In some cases, up to 40 sprinklers could operate in control mode, resulting in significant water damage.

In contrast, the ESFR sprinkler operates in a fire suppression mode. ESFRs are designed to respond quickly as the fire develops and deliver a heavy sprinkler discharge onto the fire source. One of the most attractive features of the ESFR is that it's designed to operate without inrack sprinklers.

Since there's no need for in-rack sprinklers, warehouse operators have maximum flexibility with storage height, arrangement, and materials handling. With the exception of special hazards, ESFRs will protect anything up to cartoned and uncartoned high density plastic products. ESFRs will also protect, foam plastic (such as styrofoam cups) stored to 25 ft. high in a 35 ft. high building. This applies to open frame (no solid shelves) racks and to solid pile pelletized storage. From a retrofitting standpoint, ESFRs can be costly. An ESFR retrofit likely will require major changes in piping and, depending on the local water supply, may also require a booster pump, fire pump, and tank upgrade. That's not to say that retrofitting with ESFRs, should not be considered. In some cases, the high level of protection, handling, flexibility, and lower smoke and water damage potential may make an ESFR retrofit cost-effective in the long run. For new construction on the other hand, ESFRs can be very cost-effective when they eliminate the initial cost of in rack sprinklers and the cost of removing and reinstalling in-rack sprinklers whenever racks are replaced.

Large Drop Sprinklers

Large Drop sprinklers discharge large volume of water which along with sufficient water pressure can penetrate a severe storage fire.

The ESFR Sprinkler Revolution

The uses ESFR sprinklers continued to evolve. Within a year of its introduction, the ESFR sprinkler's protection limits were extended to include even more challenging occupancies than unexpanded plastics in cardboard cartons. One of the first was aerosol storage, followed by roll paper, idle pallets, and expanded plastics.

Aerosols

Level 1 aerosol products can be protected by ESFR sprinklers up to 25 feet high in a building up to 30 feet high. These products no longer need to be stored in cut-off-areas. The more hazardous, levels 2 and 3 aerosols can be protected by ESFR sprinklers, but only at a lower storage and lower ceiling heights.

Roll Paper

Like aerosols and plastics, roll paper has long been recognized as a high hazard combustible.

ESFR sprinklers can provide effective protection for some certain arrangement for heavy and medium weight roll paper.

Pallets

Storage of idle pallets continues to be a major fire risk in warehouses. Once ignited, idle pallets arranged in stacks can burn intensely enough to threaten the structural steel of a warehouse. ESFR can provide protection to 16 ft. high storage of wood pallets. These must be stored directly on the floor not in racks.

Plastics

ESFR sprinklers at 50 psi can protect cartoned expanded and uncartoned unexpanded plastics stacks up to 25 ft. high in a 30 ft. high building. Uncartoned, expanded plastics require ESFR sprinklers at 75 psi with storage heights up to 20 ft. in a 25 ft. high building.

Much of the focus of ESFR testing remains on the various types and forms of plastics. For example, recent testing confirmed protection can be designed to 35 ft. high cartoned, unexpended plastics in a 40 ft. building if sprinkler discharge pressure is boosted to 75 psi. These ESFR applications represent a curve of growing knowledge that can change as research continues. Applicability of ESFR sprinklers continues to rely primarily on a sufficient number of tests from which to draw meaningful conclusions.

Large Drop sprinklers work in control mode. So, warehouses with Large Drop, located just on the ceiling, can still expect moderate smoke and water damage depending on the type and extent of the fire. This damage is lessened when Large Drops are used in conjunction with in-rack sprinklers.

The Large Drop's protection capabilities include anything up to cartoned, high-density or foam plastic, except special hazards, stored up to 20 feet in buildings of varying heights. Large Drop will also protect uncartoned foam plastic up to 20 feet as long as it's stored on the floor. This height is raised to 25 feet if one level of in-racks is used.

Depending on the situation, Large Drops can eliminate the need for in-rack sprinklers as long as the preceding restrictions are obeyed. In some high storage situations, the number of in-rack sprinklers required may be reduced as a result of Large Drops at the ceiling.

For existing buildings Large Drop sprinklers is very cost effective if the building owner is looking for greater storage flexibility. In situations where larger piping or booster pumps are needed, the cost will increase accordingly. For new construction, Large Drops may be a good choice for flexibility but they cannot compete with the flexibility of the ESFR in terms of storage height and fire suppression.

Standard Sprinklers

The standard sprinkler has been protecting buildings for more than a century. But, the 1960s brought many changes to warehouse practices that brought to light the limitations of the standards sprinkler, though they continued to provide low-cost protection for certain storage arrangements.

There are three types of standard sprinklers, each defined by its orifice size. The 1/2 in.

sprinkler is the oldest. The 17/32-in. was introduced in the early 1960s. The Extra Large Orifice (ELO) became available in the 1980s. Basically, the ELO is an improvement over the 17/32-in which in its day was an improvement over the 1/2-in model.

ELO sprinklers are effective where there's a low pressure, high volume water supply. ELOs may eliminate need for a booster pump in these situations.

All three of these standard sprinklers operate in control mode. Depending on the type of fire, moderate to high smoke and water damage can be expected. This will be lessened in a rack storage system if in rack sprinklers are used. Since 1/2 in. sprinklers typically require very high pressure to protect plastic, lets focus on comparing the 17/32-in. and the ELO with the ESFR and Large Drop. With the exception of special hazards, both the 1 7/32-in. or the ELO at the ceiling only can protect anything up to cartoned, high density of foam plastic storage up to 20 ft. in a 25 ft. building. Either sprinkler type at the ceiling only can protect uncartoned, high density plastic stored up to 20 ft. in a 25 ft. building.

From a cost and retrofit standpoint, for existing buildings, using ELO sprinklers may be a cost effective choice since booster pumps may not be required. And, when faced with an inadequate 1/2 in. or 17/32-in. protection scheme due to low water pressure, ELOs may be the best answer. Several factors must also be considered when looking at the cost effectiveness of standard sprinklers in new construction. For example, if a 40-ft high warehouse is being built, why standard sprinklers when the ESFR can maximise flexibility.

On the other hand, if the warehouse being constructed is higher than 40 ft., there's no choice but to use standard sprinklers with inracks, or even large Drops with inracks. In this case, the higher the storage, the higher the cost.

Finally, consider the cost associated with the removal and reinstallation of inrack sprinklers in the future when storage racks are replaced.

It should be apparent that designing fire protection is not a simple matter and requires professional assistance by knowledgeable and experience fire protection engineers.

Fire Protection for Plastic Material

Over the years, the materials used in many products have changed from more traditional wood, cloth, glass, and metal to synthetic materials. The proliferation plastics in particular is reflected in the contents of storage areas and warehouses. Storage that once consisted of glass bottles, cans, metal parts, and the like now is made up of plastic of similar materials. Even packaging is plastic.

From a fire protection standpoint plastics present a significantly greater fire hazard potential than petroleum based products. They burn faster and hotter and require more protection. Several recent multimillion dollar fires have demonstrated this need. Storage areas with sprinkler systems and other types of fire protection originally designed for less hazardous commodities than plastics.

When types of storage changes, it is critical that plant takes necessary steps to modify the fire protection accordingly.

One company that has found itself handling more and more plastics in the manufacture of electronic components and computer parts is IBM. Aware of the increased hazard that plastics present, the company began a project to reevaluate its fire protection measures and make sure its systems were up to date.

The first thing that they did was to conduct a review of the plant's manufacturing and storage locations and classify the commodities and the types of plastics used. Then, either fire protection was up graded or the amount of plastic used to package the product was reduced.

The quantity of plastics also presented a hazard, so the firm sought to limit the amount of plastics on the site. In one move it reduced the volume of plastic packaging, by using foam corners blocks instead of encapsulating the product. Reduction also is promoted by a just-in-time program—a continuous flow process that reduces the amount or inventory stored.

In addition, the level of plastics on the assembly floor at any one time is restricted. For example, the foam protective packaging, which burns rapidly and has an extremely high heat release, is stored in tractor-trailers outside the facility and brought in on conveyors from the docks when needed. Because it has a high fire hazard potential, only a four hour supply is allowed in the plant and any trash is cleared away continuously by conveyor. Combustible materials, raw goods, or finished products that are packaged in foam and cardboard also are held to a minimum. Plant fire protection personnel monitor the areas for compliance through weekly inspections and frequent spot checks.

Because operations are continually changing the company reviews its fire protection measures whenever there is a change in a manufacturing or warehousing configuration. In each case, the circumstances are examined for the least costly alternative and if justified, the fire protection is upgraded to a level suitable to plastic hazards. The firm also is considering conducting fire tests to analyze the fire behaviour of the high-density polymers in its storage areas and determine the level of protection they need.

30 Rack Storage of Materials

This chapter provides fire protection guidelines and standards for storage of Class I, II, III, and IV commodities on racks to any height. These guidelines must be applied with the understanding that commodities tested in the rack storage test program consisted, for the most part, of approximately 3½ ft. (1.1 m) cubed, closed- top, cardboard cartons containing various materials, stored on wooden pallets on standard racks having approximately 5 ft. (1.5 m) high tiers. Various storage methods and configurations exist which may alter either the fire behavior or the effectiveness of the fire protection contemplated by this data sheet. Approach these situations with careful judgement to ensure provision of adequate fire protection.

These standards do not provide fire protection guidelines for storage in structures having many levels, each less than 3½ ft. (1.1 m) high, using either open-top or closed-top containers with few, if any, flue spaces.

A. General Definitions

Aisle. An aisle is the clear space normally maintained between storage racks, or between commodities overhanging the edge of the racks, for transfer of commodities to or from racks.

Bulkhead. A bulkhead is a vertical barrier across the rack in the direction of rack loading. Bulkheads are not a substitute for in-rack sprinklers when in-rack sprinklers are recommended.

Clearance. Clearance is clear space maintained between the top of storage and ceiling sprinkler deflectors.

Commodity. A commodity is the combination of product, packaging material, container, and material handling device, upon which commodity classification is based.

Encapsulation. Encapsulation is a method of packaging consisting of a plastic sheet completely enclosing the sides and top of a pallet load containing a combustible commodity or a group of combustible commodities or combustible packages. Totally noncombustible commodities on wood pallets enclosed only by a plastic sheet as described are not considered encapsulated. The term encapsulation also applies to individual cartons which are enclosed on the top and sides in plastic, or to cartons waterproofed by coatings on the exterior surfaces. The term encapsulation does not apply to individual plastic enclosed items inside a larger non-plastic enclosed or no waterproofed container. Where there are holes or voids in the plastic or waterproof cover, on the top of the carton, which exceed more than half of the area of the top, the term encapsulation does not apply.

Flue Space. The flue spaces are the open spaces between rows of storage. Longitudinal flue spaces are perpendicular to the direction of loading. Transverse flue spaces are parallel to the direction of loading.

Horizontal Barriers. Horizontal barriers are solid barriers located on a horizontal plane within a rack covering all vertical openings in the rack, beneath which in-rack sprinklers are installed. They extend to both ends and both aisle traces of the racks, and are fitted to within 2 in. (50 mm) of any vertical rack member of other equipment which would create an opening, such as vertical in-rack sprinkler pipe drops. Their purpose is to impede the vertical propagation of fire by blocking off normally open flue spaces.

Pallets. Pallets are material handling aids which rest on rack members and support a unit load. Conventional pallets have stringers to facilitate use of material handling devices, such as lift trucks. Slave pallets (less than 20 ft^2 (1.9 m^2) are special flat bottom wood pallets captive to a material handling system.

Rack Storage Sprinklers Rack storage sprinklers are standard sprinklers equipped with an attached water shield for use as in-rack sprinklers. They can be located in the longitudinal flues, the transverse flues, or at the face of the rack. Flue sprinklers oppose vertical tire growth up through the flues. Face sprinklers are located within 18 in. (0.5 m) of the aisle face of storage, and oppose vertical fire growth on the external face of storage. They simultaneously provide a water curtain between two adjacent racks to hinder fire jump across the aisle.

Solid Shelving. Solid shelving is solid, slatted, or other types of shelves located within racks which can promote horizontal fire spread throughout the racks and obstruct sprinkler water penetration down through the racks. The area of a solid shelf is defined by minimum 6 in. (150 mm) flue spaces at all four edges of the shelf. Racks with fixed slatted shelves having uniform openings of at least 50%, with the maximum slat width between openings not exceeding 6 in. (150 mm) and with at least 3 in. (75 mm) transverse flues at rack uprights or at least every 10 ft. (3.1 m) of rack length, can be treated as racks without solid shelves. Racks with grated mesh or other type shelves having uniform openings not more than 6 in. (150 mm) apart, comprising at least 50% of overall shelf area, and with adequate flue spaces between blocks of storage on the same tier, can be treated as racks without solid shelves.

Storage Height. Storage height is measured from the floor to the topmost level of storage. For new or proposed buildings or structures, storage height is the distance from the floor to within 3 ft. (0.9 m) of ceiling sprinkler deflectors.

Storage Racks. Storage racks are any combination of vertical, horizontal and diagonal members, with or without solid shelves, that support stored material.

Racks may be fixed in place or portable. Loading may be either manual using lift trucks, stacked cranes, or hand placement, or automatic using Machine controlled storage and retrieval systems.

Single-row racks are racks with no longitudinal flue space and having a width up to 6 ft. (1.8 m) with aisles at least 3½ ft. (1.1 m) from other storage. Double-row racks are two single-row racks placed back to back, separated by a longitudinal flue space with aisles at least 3½ ft.

(1.1 m) on each side. Multiple-row racks are racks greater than 12 ft. (3.7 m) wide, or single-row or double-row racks separated by aisles less than 3½ ft. (1.1 m) wide having an overall width including flues, greater than 12 ft. (3.7 m).

Portable racks are not fixed in place. They can be moved in any direction to make the desired arrangement. Fire protection needs are determined by pile dimensions as given above for movable racks are racks on fixed rails or guides. They can be moved in one direction only in a horizontal two-dimensional plane. A moving aisle is created as abutting racks are either loaded or unloaded, then moved across the aisle to abut other racks. Rack arrangements generally result in the same protection needs as for multiple-row racks.

Tier. A tier is a vertical segment of storage, between horizontal rack members, nominally 5 ft. (1.5 m) high, but within a 3½% to 6 ft. (1.1 to 1.8 m) range. Location and recommended number of levels of in-rack sprinklers are specified in terms of tiers within the range of these dimensions.

B. Building Construction

Building construction may be conventional with storage racks placed inside the structure, or the racks themselves may form the structure. In a conventional building, the rack design is independent from the building wall and roof design. In a rack-supported structure, the racks are used to support the roof and walls.

C. Storage Arrangements

Storage arrangements vary between locations. Most arrangements can be classified as single-row, double-row, or multiple-row racks for the purpose of determining fire protection requirements. Fire protection guidelines in this data sheet are based on fire tests using materials in cardboard containers on wooden pallets stored in tiers. Other storage arrangements exist which may promote fire behavior different from that contemplated by the protection guidelines contained in the data sheet. The data sheet discusses structures using storage containers of small height inserted into the structure to form many storage levels, having essentially no flue spaces in a relatively low overall storage height.

D. Fire Tests

In 1967, concern over the increasing use of racks for storage in warehouses prompted a consortium to initiate a test program aimed at providing tire protection guidelines for these high value, high hazard storage arrangements. A series of more than 80 full scale fire tests was conducted to determine minimum fire protection needs for the various expected storage arrangements and commodities. The test program was divided into two parts: storage up through 25 ft. (7.6 m) high, and storage higher than 25 ft. (7.6 m).

In 1979, Factory Mutual recognized that storage practices not addressed by the original test program were becoming more common, and embarked on a supplementary test program to study the effects of these conditions on fire severity. Specific conditions tested were aisle storage between racks, irregular flue spaces between pallet loads, an additional pallet load of storage on the top tier of the rack, high initial sprinkler discharge pressures, and mixed storage of commodities having different fire hazards. These conditions were superimposed upon an otherwise adequately protected rack storage test array.

E. Fire Behavior and Control

The rack storage test programs provided a great deal of insight into fire behavior in rack storage.

A rack storage fire typically travels up through the flue spaces in a V-shaped pattern to the top of the rack. Fire control is achieved when operating sprinklers reduce the fire to one of no growth or of diminishing size. This is accomplished by prewetting materials surrounding the central fire core to prevent further fire spread, while allowing materials already involved to be extinguished or burn themselves out.

Many protection and storage arrangement variables exist that can influence fire behavior and fire control. The effect on fire behavior of the interplay between the many variables is complex, and not all variables have been tested in conjunction with other variables. A discussion on the advantages and disadvantages of some of these variables follows.

1. In-rack Sprinklers

The value of in-rack sprinklers in achieving fire control is fully supported by test work and loss experience. Factory Mutual has never experienced a catastrophic loss from a fire starting in rack storage where in-rack sprinklers were provided and operated properly. In-rack sprinklers are needed whenever conditions exist which can diminish or prevent ceiling sprinkler water penetration down through the racks to fire in lower tiers, or when ceiling sprinklers alone, due to hydraulic limitations, cannot provide enough water to control the fire. With insufficient water penetration, a fire may rapidly grow out of control, or it may continue in a controlled growth state until enough sprinklers have opened to deplete water supplies to a point where little or no fire control exists. This will result in a rapid increase in fire growth and heat release, with subsequent roof steel failure and burnout of the storage area. Installation of in-rack sprinklers should be encouraged whenever possible.

Increased ceiling sprinkler discharge density is not an adequate substitute for needed in-rack sprinklers.

Fire tests have shown the need for water shields over in-rack sprinklers. Fire tests have also shown the need for properly located face sprinklers for most storage higher than 25 ft. (7.6 m) and for lower storage under special conditions. They are critical to achieving fire control. Without face sprinklers where needed, an intense fire plume may develop on the aisle face of storage, promoting fire jump to adjacent racks and increased fire spread.

2. Clearance

Both fire testing and loss experience show the advantage of low clearance between ceiling sprinklers and the top of storage. Both test work and a rack storage loss survey revealed that the average number of opened sprinklers and the average damage increase with increasing clearance. A 3 ft. (0.9 m) clearance is considered the most advantageous from a fire control standpoint.

3. Flue Spaces

Adequate flue spaces are essential to allow sufficient sprinkler water penetration down through the racks. The guidelines set forth in this standard were developed through actual test work.

Flue spaces are critical when solid shelves or slave pallets are present, both to define shelf area and to allow sprinkler water penetration down through the rack.

Vertical alignment is essential. Test work in racks with solid shelves has shown that failure was due to fire spreading beneath shelves and jumping flue spaces until the ends of the ignition rack were reached. Conclusions were that fire would have continued to spread had the rack been longer.

4. Solid Shelves

Solid shelves provide smooth, often combustible, horizontal surfaces which promote rapid lateral fire spread beneath the shelves. It is difficult for ceiling sprinklers alone to control fire in racks with solid shelves, because of both the rapid fire spread through the rack, and the decreased prewetting of materials around the fire due to the shelves obstructing the sprinkler water penetration down through the rack. Fire spreading beneath solid shelves in the lower tiers encounters little resistance in the form of prewetted materials. Once the fire jumps an aisle, again there is little resistance to fire spread throughout the rack. Eventually, enough sprinklers open to deplete the water supply to a point of total ineffectiveness, resulting in a completely uncontrolled fire.

Slatted, grated, wire mesh, or other type shelves may present the same fire hazard as solid shelves, especially if commodities stored in the racks are not palletized, creating smooth shelf surfaces with essentially no openings. Fire testing has shown the need for flue spaces to define shelf areas. Where shelf areas are large enough to promote rapid fire spread, testing has shown the need for in-rack sprinklers.

5. Aisle Width

Maintenance of aisles free from storage is essential for adequate fire protection. Protection recommendations contained in this data sheet are dependent upon aisle width. Fire tests clearly indicated an increasing fire hazard with decreasing aisle widths. Since the intensity of radiant heat is approximately inversely proportional to the square of the distance from the heat source, aisle storage which effectively reduces an 8 ft. (2.4 m) aisle to 4 ft. (1.2 m) dramatically increases the radiant heat to the face of storage across the aisle. This will promote earlier fire jump across the aisle. A more severe fire will result due to the reduced effectiveness of protection and time available for prewetting commodities in the rack (or aisle storage) to which the fire jumps.

An otherwise adequately protected 19 ft. (5.8 m) high double-row rack array of Class 2 commodity, with 8 ft. (2.4 m) aisles, had 16 ft. (4.9 m) high palletized Class 2 storage placed in one aisle adjacent to a target rack, effectively reducing the aisle width to 4 ft. (1.2 m). As a result of the aisle storage and the reduced aisle width, protection was inadequate. Ceiling air temperatures were high enough for a long enough period to threaten roof steel integrity.

A similar test, using 8 ft. (2.4 m) high idle wooden pallets adjacent to the ignition rack as the aisle storage, gave similar results.

6. Storage Height

Fire testing has shown the fire hazard to increase exponentially with increasing storage height. Loss experience shows a trend towards greater damage and more sprinklers opening with increasing storage height.

An otherwise adequately protected 19 ft. (5.8 m) high double-row rack array of Class 2 commodity had one additional pallet load of material placed on the top of the uppermost storage tier. Overall storage height was 23 ft. (7 m). Test results indicated protection was inadequate and ceiling air temperatures were high enough for a long enough period to threaten roof steel integrity.

7. Mixed Commodities

Fire testing has shown that the hazard of mixed commodities within a rack storage array can approach that of the most hazardous commodity present. Two tests were run using an otherwise adequately protected 19 ft. (5.8 m) high double-row rack array of Class 2 commodity, with the top tier of storage replaced with a plastic commodity in one test, and the bottom tier of storage replaced with a plastic commodity in the other test. Both tests indicated totally inadequate protection and a very severe fire hazard.

Various commodities located randomly throughout a warehouse can present as severe a fire hazard as a storage of only the most hazardous commodities.

8. Storage Arrangement

Trends in materials handling are towards more efficient, automated storage and retrieval, with emphasis on maximum use of available storage space. Two storage practices which have evolved from this trend, and which may promote tire behavior not specifically contemplated by the protection guidelines contained in this data sheet, are the use of open-top containers and multi-level storage and retrieval systems. Often, these storage arrangements contain highly concentrated values in relatively small areas.

Open-top containers can collect water from sprinklers, preventing penetration down through the racks to fire in lower levels. An otherwise adequately protected 19 ft. (5.8 m) high double-row rack array of Class 2 commodity, with the tops of all containers on the top tier, and the central eight containers on the third tier, cut open and removed, resulted in a higher water demand and approximately twice as much damage as in a test with closed-top containers. The fire was stopped before reaching the ends of the ignition rack, but jumped both aisles and spread to the outermost portions of both target racks. Had the array been larger, continued fire spread across additional aisles could have been expected. Fire severity was attributed to the open-top containers collecting water which would otherwise percolate to lower levels of storage. The fire hazard of storage arrangements significantly different than those tested to develop the protection guidelines contained in this data sheet depends to a great extent upon factors which could adversely affect fire behavior and control. Any evaluation of these storage arrangements should take into account all factors present.

9. Sprinkler Temperature Rating

In tests with storage up to 25 ft. (7.6 m) high with both ceiling and in-rack sprinklers, the first ceiling sprinkler operated prior to the first in-rack sprinkler. Tests with ceiling sprinklers only showed fewer 286°F (141°C) than 165°F (74°C) sprinklers operated. Results of these tests conclude that 286°F (141°C) ceiling sprinklers are preferred and require a lesser water demand than 165°F (74°C) ceiling sprinklers. Tests in storage higher than 25 ft. (7.6 m) all had in-rack sprinklers, with the first in-rack sprinkler operating prior to any ceiling sprinklers. Results showed fewer 165°F (74°C) than 286°F (141°C) ceiling sprinklers operated; hence, the lesser ceiling sprinkler water demand for 165°F (74°C) sprinklers.

Loss Experience

Illustrative Losses

Inadequate Sprinkler Protection Unable to Control Fire Involving Aisle Storage and Solid-Shelved Racks.

Clothing and shoes were stored in single row racks to 22 to 28 ft. (6.7 to 8.9 m) high. Idle pallets up to 12 ft. (3.7 m) high were stored in 6 ft. (1.8 m) wide aisles between racks. Solid shelves were present in the lower two tiers to form picking bins. Needed in-rack sprinklers were not provided.

A fire starting in this area overtaxed the ceiling sprinkler system. Roof collapse occurred within 20 minutes of fire discovery. Approximately 200,000 ft^2 (18,600 m^2) of this building was destroyed. Property damage $21,000,000.

Fire in High Rack Storage Controlled by In-Rack Sprinklers. A fire in clothing in cartons on 16-tier, 39-ft. (11.9 m) high racks was well controlled by four in-rack sprinklers and one ceiling sprinkler. In-rack sprinklers were provided at four levels. Two small hoses were needed to complete extinguishment. Property damage $145,000. Business interruption $2,000.

Lack of In-rack Sprinklers for Racks with Solid Shelves Results in Extensive Fire Damage. Stuffed furniture was stored on cantilever racks to 15 ft. (4.7 m) high. Racks were 8 ft. (2.4 m) wide with 56 ft.' (5.2 m^2) solid plywood shelves. Needed in-rack sprinklers were not provided. A fire starting in the rack storage quickly overtaxed ceiling sprinklers with initial roof collapse occurring within 20 minutes of fire discovery. Approximately half of the 202 by 405 ft. (60 by 120 m) building's roof collapsed and storage burned. Roof collapse ahead of the fire apparently hindered further fire spread by blocking flames from reaching uninvolved areas. Property damage $3,400,000.

Open-top Containers in Racks Interfere with Sprinkler Water Penetration Resulting in Uncontrolled Fire. Synthetic fiber socks were stored in cardboard "tote boxes" in racks 16 ft. (4.9 m) high. There were eight levels of boxes supported on metal angles in each rack. A fire starting in or near the racks destroyed a 390 by 530 ft. (120 by 160 m) portion of this plant. Water from ceiling sprinklers collecting in boxes on the top levels apparently prevented adequate sprinkler water penetration down through the racks to control the fire. In-rack sprinklers were provided when the plant was rebuilt. Property damage $14,000,000.

General

The premier trend evolving from a survey of 147 rack storage losses is the overwhelming success in fire control which can be attributed to in-rack sprinklers. To date, Factory Mutual has not experienced a catastrophic loss in rack storage when in-rack sprinklers were installed and in service at the time of the fire.

Other major trends derived from the same survey are:

1. Both damage and the number of sprinklers opened increase with increasing storage height.
2. Both damage and the number of sprinklers opened increase with increasing clearance between ceiling sprinklers and
3. The number of sprinklers opened increases with increasing commodity hazard.

Protection defects were identified in all rack storage losses where the fire was determined to be uncontrolled.

Recommendations

Building Construction

a. Design. Construct buildings in accordance with existing standards. Design rack-supported structures taking into consideration the effects of wind and snow, rack loads, seismic conditions, and the additional load created by Collection or absorption of fire protect ion water, the weight of water-filled sprinkler piping, and any other loads to which the structure may be exposed.

b. Steel Protection. When sprinkler systems are installed in accordance with this standard, fire protection of roof steel is not necessary.

When sprinkler systems using both ceiling sprinklers and in-rack sprinklers are installed in accordance with this standard, fire protection of steel building columns is necessary. When storage height exceeds 15 ft. (4.6 m), and ceiling sprinkler s only are installed, provide fire protection by one of the following methods f or all types steel building columns located (wholly or partially) within the racks or for vertical rack members that support the structures.

i. Sidewall sprinkler, at the 15 ft. (4.6 m) elevation, pointed towards one side of the steel column or vertical rack member.

ii. One hour fireproofing.

iii. For storage heights up through 20 ft. (6.1 m), provision of ceiling sprinkler density for a minimum of 2000 ft² (186 m²) with 286°F (141°C) or 165°F (74°C) rated sprinklers as follows:

Commodity Class	4 ft. (1.2 m) Aisles, gpm/ft² (mm/min)	8 ft. (2.4 m) Aisles, gpm/ft. (mm/min)
I	0.37 (15. 1)	
II	0.44 (1 7.9)	
III	0.49 (20.0)	

c. Heat and Smoke Venting. Factory Mutual recommended protection is based on roof vents and draft curtains not being provided. Fire tests have not shown automatic vents to be cost effective and they may even increase sprinkler water demand. Hence, permanent heat and smoke vents, if any, should be arranged for manual operation. Smoke removal during mop-up operations can frequently be achieved through eaveline windows, doors, monitors, non-automatic exhaust systems (gravity or mechanical), or manually operated heat and smoke vents. Fire departments can cut holes in steel or wood roofs and also use their smoke exhausters.

Storage Arrangements

Most storage arrangements can be classified as single-row, double-row, or multiple-row racks for the purpose of determining fire protection requirements.

a. Rack Loading. Do not load racks beyond their design capacity. When designing a rack structure, consider the actual loads being supported, and additionally, the loads created by water-filled in-rack sprinkler piping and by sprinkler water collection or absorption by the commodities being stored. Added loads due to sprinkler water collection or absorption will vary depending upon the type of material stored and the storage method. In locations where seismic resistance or buildings is necessary, design rack structures for those seismic conditions.

   Adequate design and maintenance of rack structures is essential to prevent rack collapse and possible breakage of in-rack sprinkler piping. Carefully evaluate and correct, if necessary, any conditions which put into question the rack structure integrity.

b. Flue Spaces. Flue spaces are necessary to allow sprinkler water penetration down through the racks. Without sufficient water penetration to burning commodities stored on the lower tiers of the racks, fire control may not be achieved.

   Rack storage up to and including 25 ft. (7.6 m) and without solid shelves does not require longitudinal flue space. Maintain at least 3 in. (75 mm) transverse flue space between loads or at rack uprights to allow sufficient sprinkler water penetration. Random variations in the width of the flue spaces or in their vertical alignment are permissible.

   Rack storage higher than 25 ft. (7.6 m) and without solid shelves requires both transverse and longitudinal flue spaces (except in single-row racks). Maintain at least 3 in. (75 mm) longitudinal flue space. Maintain at least 3 in. (75 mm) transverse flue spaces between loads or at rack uprights. Maintain vertical alignment of flue spaces as closely as possible.

   In single-row, double-row, and multiple-row racks any height with solid shelves, maintain minimum 6 in. longitudinal and transverse flue spaces between shelf sections, thereby defining shelf area and permitting sprinkler water penetration. Maintain vertical alignment of the flue spaces as closely as possible.

c. Pallets. Plastic pallets may be used in rack storage where in-rack sprinklers are not required. Do not use plastic pallets in rack storage with in-rack sprinklers.

   The softening of the plastic pallets during a fire could result in shifting of loads and the breaking of in-rack sprinkler piping.

Fire Protection

General

Install wet-pipe sprinkler systems whenever possible. Use preaction sprinkler systems only when it is impractical to provide building heat. Avoid the use of dry-pipe systems whenever possible. When dry systems are unavoidable, pipe schedule systems are preferred and should normally be limited to 500 gal (1893 dm^3) or less capacity and be equipped with a quick opening device. For other than pipe schedule systems or pipe schedule systems of larger or unknown capacity, water delivery time to the inspector's test connection should be 60 seconds or less, starting at normal air pressure on the system and with the quick opening device in operation. Dry-pipe in-rack sprinkler systems should not be used in locations normally exposed to freezing temperatures.

Protection consisting of ceiling sprinklers and in-rack sprinklers is the preferred method to use. This type of protection results in much less water and fire damage during a fire and assures greater probability of successful control. Ceiling sprinklers alone, where allowed for rack storage protection, is a less desirable alternative.

A. Ceiling Sprinklers

1. Type Standard 17/32 in. (14 mm) orifice, 286°F (141°C) rated sprinklers are preferred, except that either 165°F (74°C) or 286°F (141°C) sprinklers may be used for storage higher than 25 ft. (7.6 m). Standard 1/2 in. (13 mm) orifice sprinklers may be used. On-off type sprinklers should not be used to protect rack storage.
   In some cases, rack storage may be protected using large- drop sprinklers. See Data Sheet 2-7 for installation guidelines.
   Extend the installation of 286°F (141°C) rated sprinklers beyond the rack storage being protected in accordance with the following table.
2. Spacing. Do not exceed the maximum sprinkler spacing recommended except that sprinkler spacing may exceed 100 ft.^{2} (9.3 m^2), but should not exceed 130 ft^2 (12.1 m^2), in systems hydraulically designed for densities below 0.25 gal/min/ft^2 (10.2 mm/min).
   Minimum spacing should be 50 ft^2 (4.7 m^2) for 1/2 in. (13 mm) orifice sprinklers, and 70 ft^2 (6.5 m^2) for 17/32 in. (14 mm) orifice sprinklers.
3. Design. Ceiling sprinkler discharge density should not exceed 0.60 gal/min/ft^2 (24.4 mm/min) nor be less than 0.15 gal/min/ft^2 (6.1 mm/min) after any density adjustments have been applied. The maximum sprinkler discharge pressure at the calculation starting point for hydraulically designed systems should not exceed 60 psi (4.1 bar, 414 kPa).
   Ceiling sprinkler design area should not exceed 6000 ft^2 (560 m^2) nor be less than 2000 ft^2 (186 m^2) after any area adjustments have been applied.

In buildings occupied in part by rack storage, extend the sprinkler system design 15 ft. (4.6 m) beyond the area occupied by the rack storage.

Determine ceiling sprinkler water demand from the applicable sections of this data sheet for storage up through 25 ft. (7.6 m) high, and for storage higher than 25 ft. (7.6 m).

For storage up through 25 ft. (7.6 m) high, ceiling sprinkler design areas for 165°F and 212°F (71°C and 100°C) rated sprinklers should be 170% and 135%, respectively, of that for 286°F (141°C) rated sprinklers.

For storage higher than 25 ft. (7.6 m), ceiling sprinkler design density for 212°F (100°C) rated sprinklers should be midway between that for 165 and 286°F (74°C and 141°C) rated sprinklers.

B. In-rack Sprinklers

1. Type. Use rack storage sprinklers with either a t/2 or 17/32 in. (13 or 14 mm) orifice. When fixed solid shelves or horizontal barriers shield in-rack sprinklers from prewetting by sprinklers above, sprinklers without water shields may be used beneath the shelves or barriers. On-off type sprinklers should not be used. Sprinklers should be rated at 165°F (70°C) or less, except that higher temperature rated sprinklers may be used near unit heaters. Preferably such heaters should be relocated.
2. Horizontal Spacing. Stagger sprinklers in single-row and double-row racks between adjacent racks. Stagger sprinklers in multiple-row racks on adjacent branch lines. One line of sprinklers is adequate for single-row racks and double-row racks up to 9 ft. (2.7 m) wide. Provide two lines of sprinklers arranged as face sprinklers and staggered horizontally for double-row racks over 9 ft. (2.7 m) up to 12 ft. (3.7 m) wide.

   For non-encapsulated storage in single-row and double-row racks up through 25 ft. (7.6 m) high, maximum horizontal spacing of sprinklers on branch lines should not exceed that shown in the following table.

   Maximum Spacing on Branch Lines

Aisle Width	Commodity Class		
	I & II	III	IV
8 ft. (2.4 m)	12 ft. (3.7 m)	12 ft. (3.7 m)	8 ft. (2.4 m)
4 ft. (2.4 m)	12 ft. (3.7 m)	8 ft. (2.4 m)	8 ft. (2.4 m)

   For encapsulated storage in single-row and double-row racks up through 25 ft. (7.6 m) high, maximum horizontal spacing of sprinklers on branch lines should not exceed 8 ft. (2.4 m).

   For non-encapsulated storage in multiple-row racks up through 25 ft. (7.6 m) high, sprinkler spacings on branch lines and areas should not exceed 100 ft² for class I, II, III and 80 ft² for class IV.

   For encapsulated storage in multiple- row racks up through 25 ft. (7.6 m) high, spacing of sprinklers on branch lines should not exceed 8 ft. (2.4 m) and area per sprinkler should not exceed 80 ft² (7 4 m²).

   For storage up to 25 ft. (7.6 m) high, when it is necessary to locate sprinklers to prevent mechanical damage to sprinklers or piping, the above recommended 8 ft. (2.4 m) sprinkler spacings can be increased to the same spacings between rack uprights, but should not exceed 10 ft. (3.1 m). Also, in multiple-row racks, maximum areas per sprinkler should not exceed those recommended above.

For storage higher than 25 ft. (7.6 m), space in-rack sprinklers as indicated in Tables 4.7 and 4.8.

3. Vertical Spacing. When only one level of in-rack sprinklers are provided for storage up through 25 ft. (7.6 m) in single-row, double-row and multiple-row racks, locate the in-rack sprinklers at a tier nearest to 1/2 to 2/3 of the storage height. When two levels are provided, locate the in-rack sprinklers at tiers nearest 1/4 to 1/3, and 2/3 to 3/4 of the storage height. When more than two levels are installed, locate in-rack sprinklers at tiers so that approximately equal heights of storage are protected by each level of in-rack sprinklers and ceiling sprinklers.
4. Design. Sprinklers in racks up through 25 ft. (7.6 m) high should discharge no less than 22 gal/min (84 dm^3/min) each, and sprinklers in racks higher than 25 ft. (7.6 m) should discharge no less than 30 gal/min (114 dm^3/min) each, irrespective of the distance between sprinklers on or between lines, based on the simultaneous operation of the most hydraulically remote:
   a. 6 sprinklers when only one level is installed in racks with Class I, II, or III commodity.
   b. 8 sprinklers when only one level is installed in racks with Class IV commodity.
   c. 10 sprinklers, five on each of two most hydraulically remote levels, when more than one level is installed in racks with Class I, II and III commodities.
   d. 14 sprinklers, seven on each of two most hydraulically remote levels, when more than one level is installed in racks with Class IV commodity.

   Balance the in-rack sprinkler system water demand with the ceiling sprinkler water demand at the point where the two systems are connected.

   The number of sprinklers on a branch line and the pipe size are restricted only by hydraulic calculation, and not by any piping schedule, except that piping less than 1 inch (25 mm) nominal diameter should not be used.
5. System Size. The floor area protected by a single system of in-rack sprinklers should not exceed 40,000 ft^2 (3720 m^2), regardless of the number of in-rack sprinkler levels.
6. Location. Locate in-rack sprinklers so that sprinklers are protected from mechanical damage without completely obstructing sprinkler discharge pattern. Locate sprinklers as far as possible above the top of a tier of storage without going above the bottom of the next tier of storage.

   Sprinklers in the longitudinal flue may be located behind horizontal rack members or behind vertical rack uprights, but not behind both at once. In order to prevent mechanical damage, it may be practical to place the in-rack sprinkler piping behind horizontal members with risers or drops feeding sprinklers a correct vertical location behind rack uprights.

   Locate face sprinklers no more than 18 in. (0.5 m) from the aisle face, arranged so that sprinkler discharge is not obstructed from reaching the aisle face of the storage. Sprinklers located behind rack uprights should be at least 3 in. (75 mm) from the uprights.

   Prior to any in-rack sprinkler installation, the proposed sprinkler locations should be checked to ensure that both adequate protection against mechanical damage and proper sprinkler discharge will be achieved.

7. Control valves. Control in-rack sprinklers by separate indicating valves independent from ceiling sprinklers. Such valves are not necessary for snail in-rack sprinkler installations of less than 20 sprinklers.
   Provide separate drains for in-rack sprinkler installations. Locate drains at low points of in-rack sprinkler piping to prevent any trapped water in piping should drainage of the system become necessary. Drains should terminate no more than 6 ft. (1.8 m) above floor or platform.
8. Piping. To minimize the possibility of in-rack sprinkler pipe breakage due to deflection, piping should preferably be either steel joined by mechanical groove couplings or welding, or copper joined by brazing. Malleable iron fittings are acceptable.
9. Alarms and Test Connections. Provide a separate waterflow alarm on in-rack sprinkler systems with more than 20 sprinklers. Provide an inspector's test connection discharging to a safe area on all in-rack sprinkler systems equipped with a waterflow alarm.

C. Hose Demands and Connections

1. Hose Demand. Allow a total of at least 500 gal/min (1900 dm^3/min) for hose streams, of which at least 100 gal/min (380 dm^3/min) should be for inside hose streams. Add hose stream allowances to the sprinkler demand at the point of connection.
2. Hose Connections. Provide permanent small hose lines 1½ in. (38 mm) not exceeding 100 ft. (30.5 m) in length, capable of reaching all rack storage areas to aid in firefighting and for mop-up operations. Small hose may be supplied from any of the following:
   a. A separate piping system for small hose stations
   b. Valved hose connections on sprinkler riser where such connections are made upstream from all sprinkler control valves
   c. Adjacent sprinkler Systems
   d. Ceiling sprinklers in the protected area when separately controlled in-rack sprinklers are provided.

D. Duration of Water Supplies

Water supplies should be capable of providing combined sprinkler and hose stream demands at adequate pressure for a minimum of 1½ hours for Class I, II, and III commodities, and 2 hours for Class IV commodities.

E. Preaction Sprinkler Systems

1. Design. Use only approved preaction sprinkler systems. Determine water demands for preaction sprinkler systems on the same basis as for wet pipe sprinkler systems.
2. Detector Type. Use only approved heat-accuated detectors on single-zone circuitry. They may be fixed, rate-of-rise, or combination fixed temperature/rate-of-rise type.

3. Detector Locations.
   a. Ceiling Sprinklers Only. Locate detectors at the ceiling on a spacing not exceeding allowable ceiling sprinkler spacing.
   b. Ceiling and In-rack Sprinklers. Locate detectors at the ceiling as recommended above. Locate detectors in the racks in the rack flues at the same vertical and horizontal spacing as in-rack sprinklers.

      For single-row and double-row racks, only one line of detectors is needed at each in-rack sprinkler level. Locate these detectors in the transverse flue in single-row racks, and in the longitudinal flue in double-row racks.

      For multiple-row racks, one line of detectors is needed for each in-rack sprinkler line at each in-rack sprinkler level. Locate these detectors in either the transverse or longitudinal flues.

F. Storage up through 25 ft. (7.6 m) High

General

Use Tables to determine ceiling sprinkler discharge density, ceiling sprinkler operating area, and needed in-rack sprinkler levels. These tables are based upon the use of 286°F (141°C) ceiling sprinklers. Select the appropriate table based on the particular rack storage arrangement involved, as follows:

- Rack Storage up through 12 ft. (3.7 m) High with 286°F (141°C) Ceiling Sprinklers.
- Single-Row and Double-Row Racks over 12 ft. (3.7 m) up through 25 ft. (7.6 m) High, without Solid Shelves.
- Single-Row and Double-Row Racks over 12 ft. (3.7 m) up through 25 ft. (7.6 m) High, with Solid Shelves Greater than 20 ft^2 (1.9 m^2).
- Multiple-Row Racks over 12 ft. (3.7 m) up through 25 ft. (7.6 m) High, without Solid Shelves.
- Multiple-Row Racks over 12 ft. (3.7 m) up through 25 ft. (7.6 m) High, with Solid Shelves Greater than 20 ft^2 (1.9 m^2).

Obtain the ceiling sprinkler design density and operating area by moving from left to right through the appropriate table and applying adjustments to ceiling sprinkler discharge density and operating area as indicated.

1. Adjustments to Recommended Protection
   a. Aisle Width. In Tables for single-row and double-row racks, use the 4 ft. (1.2 m) aisle design curves for aisles 3½ to 4 ft. (1.1 to 1.2 m) wide, and use the 8 ft. (2.4 m) aisle design curves for aisles 8 ft. (2.4 m) wide and greater. For aisle widths between 4 and 8 ft. (1.2 and 2.4 m), a direct linear interpolation between curves may be made.

      Protect single-row and double-row racks with aisles narrower than 3½ ft. (1.1 m) as multiple-row racks.

      Aisles are hot needed where racks abut noncombustible building walls.
   b. Storage Height. Application of the adjustment to the ceiling sprinkler density obtained from the sprinkler system design curves is allowed when indicated by Tables.

c. Slave Pallets and Solid Shelves less than 20 ft^2 (1.9 m^2). Apply a 20% ceiling density increase when indicated in Tables.

d. In-rack Sprinklers. Adjustment to ceiling sprinkler density is allowed in as indicated by Tables when extra levels of in-rack sprinklers are provided.

e. Clearance from Top of Storage to Sprinklers. Recommended protection is based on a nominal 10 ft. (3.1 m) clearance between ceiling sprinklers and the top of storage. When clearance between ceiling sprinklers and the top of storage is 4½; ft. (1.4 m) or less, adjustment to ceiling sprinkler operating area is allowed when indicated in Tables.

   When clearance between ceiling sprinklers and the top of storage exceeds 20 ft. (6.1 m), install horizontal barriers extending to the fall length and width of the racks 3 ft. (0.9 m) above the top of storage, and provide one level of sprinklers beneath the barriers. The sprinkler system under the barrier should be designed the same as a ceiling sprinkler system for the hazard being protected. Additionally, the sprinkler system at the roof should provide at least 0.15 gal/min/ft^2 (6.1 mm/min) over 2000 ft^2 (186 m^2) for Class I commodities, and 0.20 gal/min/ft^2 (8.2 mm/min) Over 2000 ft^2 (186 m^2) for Class II, III, and IV commodities. Water demands for the sprinkler systems at the ceiling and beneath the barriers should be available concurrently.

f. Dry Pipe Systems. When dry-pipe systems are used, add 30% to the ceiling sprinkler operating area.

Storage Higher than 25 ft. (7.6 m)

General

Storage higher than 25 ft. (7.6 m) presents a severe challenge to fire protection systems. In-rack, sprinklers are needed for all storage arrangements.

1. Single-Row Racks. Protect single-row racks without solid shelves in Accordance with the Tables and provide ceiling sprinkler protection. Treat single-row racks wider than 6 ft. (1.8 m) as double- row racks.

2. Double-row Racks. Protect double-row racks without solid shelves in accordance with the Tables.

3. Multiple-row Racks. Protect multiple-row racks without solid shelves in accordance with the Tables.

   a. Solid Shelves. When solid shelves exceed 50 ft^2 (4.7 m^2) in area, provide in-rack sprinklers for single-row, double-row, and multiple-row racks in the flues of each unsprinklered tier in addition to the in-rack sprinklers. Provide ceiling sprinkler protection of 0.15 gal/min/ft^2 (6.1 mm/min) over 2000 ft^2 (186 m^2) when there is no storage above the top level of in-rack sprinklers, or the same protection as for racks without solid shelves when there is storage above the top level of in-rack sprinklers.

      When solid shelves are 20 through 50 ft. (1.9 through 4.7 m^2) in area, provide in-rack sprinklers in according to row racks for single-row racks and for multiple-row racks.

Provide ceiling sprinkler protection of 0.15 gal/min/ft^2 (6.1 mm/min) over 2000 ft^2 (186 m^2) when there is no storage above the top level of in-rack sprinklers, or the same protection as for racks without solid shelves when there is storage above the top level of in-rack sprinklers.

When solid shelves are less than 20 ft^2 (1.9 m^2), no adjustment to recommended protection is needed.

b. Excessive clearance. When clearance between ceiling sprinklers and the top of storage exceeds 20 ft. (6.1 m), install horizontal barriers extending to the full length and width of the racks 3 ft. (0.9 m) above the top of storage, and provide one level of sprinklers beneath the barriers.

Provide ceiling sprinkler protection, for single-row and double-row racks and for multiple-row racks, as referenced in the tables. Design sprinklers beneath the barriers as part of the in-rack sprinkler system.

For single-row and double-row racks, an alternative to horizontal barriers is to provide additional in-rack sprinklers above the top of storage. Provide ceiling sprinkler protection the same as for racks without solid shelves and with clearance less than 20 ft. (6.1 m).

c. Dry-Pipe Systems. When dry-pipe systems are used, add 30% to the ceiling sprinkler operating area.

H. Other Storage Arrangements

Rack storage using open-top combustible containers and rack storage in movable racks are two storage arrangements which need special consideration.

Open-top containers can collect sprinkler water and prevent water penetration to fire in lower tiers.

Movable racks are designed to use as much of the available storage space as possible. Usually racks are on tracks and are butted against one another. Loading and unloading takes place in a common aisle. Loaded racks are stored on one side of the aisle, and empty racks are stored on the other side of the aisle.

Apply the following protection guide aisle on the basis of the worst hazard which could be expected.

1. Open-top Combustible Containers. Protect rack for double-storage with open-top containers the same as racks with 50 ft^2 (4.7 m^2) solid shelves, except that at least one level of in-rack sprinklers is needed for all storage higher than 8 ft. (2.4 m) with more than one level of containers. If containers are noncombustible, these guidelines do not apply. In order for a container to be considered noncombustible, liners, dividers, or contents should not protrude above the noncombustible sides.
2. Movable Racks. Protect rack storage in movable racks the should be supplied through flexible in-rack sprinkler system connections using approved joints.

31 Centrifugal Fire Pumps

This chapter deals with the selection and installation of fire pumps supplying water for private fire protection.

The concerns would include water supplies, suction, discharge and auxiliary equipment; power supplies, electric drive and control; internal combustion engine drive and control; steam turbine drive and control; acceptance tests, and operation and maintenance.

The purpose is to provide a reasonable degree of protection for life and property from fire through installation requirements for centrifugal fire pumps based upon sound engineering principles, test data and field experience. It includes single stage and multistage pumps of horizontal or Vertical shaft design. Standards are established for the design, installation and maintenance of these pumps, pump drivers and associated equipment. The standard continues the excellent record that has been established by centrifugal pump installations and meet the needs of changing technology. Nothing in this standard is intended to restrict new technologies or alternate arrangements provided the level of safety prescribed by the standard is not lowered.

Pumps other than those specified in this standard, and having different design features, may be installed when such pumps are UL listed and approved by the authorities having jurisdiction.

Centrifugal fire pumps shall not be purchased until conditions under which they are to be installed and used have been determined. Each pump, driver, controlling equipment, power supply and arrangement, and water supply shall be approved by the authority having jurisdiction for the specific field conditions.

The pump manufacturer shall be given complete information concerning the water and power supply characteristics.

Prior to shipment of new equipment or alteration of existing equipment, a complete plan and detailed data describing pump, driver, controller, power supply, fittings, suction and discharge connections, and water supply conditions shall be prepared by the engineer or contractor for approval.

The pump, driver, and controller shall be purchased under unit contract stipulating compliance with this standard and satisfactory performance of the entire unit when installed. The pump manufacturer shall be responsible for the proper operation of the complete unit assembly as indicated by field tests.

Certified shop test curves showing head-capacity, efficiency, and brake-horsepower of the pump shall be furnished by the manufacturer.

Where existing pump installations meet the provisions of the standard in effect at the time of purchase, they may remain in use provided they do not constitute a to life or property.

Controllers

Controller means the cabinet, motor starter, circuit breaker and disconnect switch, and other control devices for the control of electric motors and internal combustion engine driven pumps.

Isolating means is a switch intended for isolating an electric circuit from its source of power.

Disconnecting means is a device, group of devices, or other arrangement (such as a circuit breaker or disconnecting switches) whereby the conductors of a circuit can be disconnected from their source of supply.

Electric Motors

Electric Motors are classified according to mechanical protection and methods of cooling.

Open Dripproof means the general-purpose motor will meet the needs of most installations where operating conditions are relatively clean and dry.

Splashproof means the motor will be protected when a plant must be washed down. The motor will be safeguarded against all falling or splashing liquids — rain, snow, sleet, etc.

Totally Enclosed Fan Cooled means the motor is protected against dusts, mist, oil, fog. Inner frame protects vital parts of the motor, seals out harmful matter.

Wound-Rotor Open Construction means the motor is ideal for applications requiring low starting current with high starting torque, reversing or adjusting speed.

Wound-Rotor Splashproof means the motor has the same electrical characteristics as immediately above and, in addition, is adequately protected against falling and splashing liquids.

Electric Supply

Eire Pump Branch Circuit means that portion of the wiring system between the motor and the final over current device protecting the circuit and the motor.

Feeder means the circuit conductors between the service equipment, or the generator switchboard of an isolated plant, and the branch-circuit over current device.

Service means the conductors and equipment for delivering energy from the electricity supply system to the wiring system of the premises served.

Service Equipment means the necessary equipment, usually consisting of a circuit breaker or switch and fuses, and their accessories, located near the pint of entrance of supply conductors to a building or other structure, or an otherwise defined area, and intended to constitute the main control and means of cutoff of the supply.

Engines

Internal Combustion Engine means any engine in which the working medium consists of the products of combustion of the air and fuel supplied. This combustion is usually effected within the working cylinder but may take place in an external chamber.

Diesel Engine means an internal combustion engine in which the fuel is ignited entirely by the heat resulting from the compression of the air supplied for combustion. The oil- diesel engine, which operates on fuel oil injected after compression is practically completed, is the type usually used as a fire pump driver.

Spark Ignited Engine means an internal combustion engine in which the fuel-air mixture is ignited by an electric discharge from a spark plug.

Head

Head is a quantity used to express a form of the energy content of water per unit weight of the water. In terms of foot-pounds (m-kg) of energy per pound (kg) of water, all head quantities have the dimensions of feet (m) of water. All pressure readings are converted into feet (m) of water being pumped. The Datum Elevation is defined as follows: For horizontal units it is the centerline of the pump shaft. For vertical double suction pumps, it is the impeller discharge horizontal centerline.

The unit for measuring head shall be the foot (m). The relation between a pressure expressed in pounds per square inch (bars) and a pressure expressed in feet (m) of head is

$$\text{Head in Feet} = \frac{\text{Pressure in psi}}{0.433}$$

$$\text{Head in Meters} = \frac{\text{Pressure in bars}}{0.098}$$

The velocity head shall be figured from the average velocity (v), obtained by dividing the flow in cubic feet per second (m^3/s) by the actual area of pipe cross section in square feet (m^2) and determined at the point of the gage connection.

Velocity head is expressed by the formula:

$$h\,v = \frac{v^2}{2g}$$

Where g = the acceleration due to gravity and is 32.17 feet per second (9.807 m/s^2) at sea level and 45 degrees latitude. v = Velocity in the pipe in feet per second (m/s)

Flooded Suction means that the water flows from an atmospheric vented source to the pump without the average pressure at the pump inlet flange dropping below atmospheric pressure with the pump operating at 150 percent of its rated capacity.

Total Suction Lift (Symbol h. s). Suction lift exists where the total suction head is below atmospheric pressure. Total suction lift, as determined on test, is the reading of a liquid manometer at the suction nozzle of the pump, converted to feet of liquid, and referred to datum, minus the velocity head at the point of gage attachment.

Total Suction Head. (Symbol H. s). Suction head exists when the total suction head is above atmospheric pressure. Total suction head, as determined on test, is the reading of a gage at the suction of the pump converted to feet of liquid and referred to datum, plus the velocity head at the point of gage attachment.

Total Discharge Head. (Symbol H. d). Total discharge head is the reading of a pressure gage at the discharge of the pump, converted to feet of liquid and referred to datum plus the velocity head at the point of gage attachment.

Total Head (Symbol H). Horizontal Pumps. Total head is the measure of the work increase per pound (kg) of liquid, imparted to the liquid by the pump, and is therefore the algebraic difference between the total discharge head and the total suction head. Total head as determined on test where suction lift exists is the sum of the total discharge head and total suction lift. Where the positive suction head exits, the total head is the total discharge head minus the total suction head.

Total Head (Symbol H_). Vertical Turbine Pumps. Total head is the distance from the water level to the center of the discharge gage, plus the discharge pressure gage reading measure just beyond the discharge elbow and referred to datum, plus the velocity head at point of gag attachment.

Total Rated Head means the total head, defined above, developed at rated capacity and rated speed for either a horizontal splitcase or a vertical shaft turbine-type pump.

Net Positive Suction Head — (Symbol h. sv). The net positive suction head is the total suction head in feet (m) of liquid absolute determined at the suction nozzle and referred to datum less the vapor pressure of the liquid in feet (m) absolute.

Pumps

A booster pump is a fire pump taking suction from a public main.

Centrifugal Pump, means a pump in which the pressure is developed principally by the action of centrifugal force.

End Suction Pump means a pump having its suction nozzle on the opposite side of the casing from the stuffing box and having the face of the suction nozzle perpendicular to the longitudinal axis of the shaft.

Fire Pump Unit means an assembled unit consisting of a fire pump, driver, controller, and accessories.

Horizontal Pump means a pump with the shaft normally in a horizontal position.

Horizontal Split-case Pump means a centrifugal pump characterized by a housing which is split parallel to the shaft.

Vertical Shaft Turbine Pump means a centrifugal pump with one or more impellers discharging into one or more bowls and a vertical educator or column pipe used to connect to bowl(s) to the discharge head on which the pump driver is mounted.

Maximum Pump Brake Horsepower means the maximum brake horsepower required to drive the pump at rated speed. The pump manufacturer determines this by shop test under expected suction and discharge conditions. Actual field conditions may vary from shop conditions.

Water Supply for Vertical Shaft Turbine-type Pump.

Aquifer means an underground formation that contains sufficient saturated permeable material to yield significant quantities of water.

Aquifer Performance Analysis means a test designed to determine the amount of underground water available in a given field and proper well spacing to avoid interference in the field. Basically, test results provide information concerning determinations of transmissibility and storage coefficient (available volume of water) of the aquifer.

Wet Pit means a timber, concrete or masonry enclosure having a screened inlet kept partially filled with water by an open body of water such as a pond, lake, or stream.

Ground Water means that water which is available from a well driven into water-bearing subsurface strata (aquifer).

Static Water Level means the level, with respect to the pump, of the body of water from which the pump is taking suction, when the pump is in operation. For vertical shaft turbine-type pumps the distance to the water level is measured vertically from the horizontal center line of the discharge head or tee.

Pumping Water Level is the level, with respect to the pump, of the body of water from which the pump is taking suction, when the pump is in operation.

Draw-Down is the vertical difference between the pumping water level and the static, water level.

Pressure Maintenance Jockey Pumps

Pressure maintenance jockey pumps should be used when it is desirable to maintain a uniform or relatively high pressure on the fire protection system. A jockey pump should be sized to make up the allowable leakage rate within 10 minutes or 1 gpm (3.8 L/min), whichever is larger.

Pressure maintenance pumps shall have rated capacities not less than any normal leakage rate. They shall have discharge pressure sufficient to maintain desired fire protection system pressure.

A check valve shall be installed in the discharge pipe.

Indicating butterfly or gate valves shall be installed in such places as needed to make the pump, check valve, and other miscellaneous fittings accessible for repair.

Where a centrifugal-type pressure maintenance pump has a shut-off pressure exceeding the working pressure rating of the fire protection equipment, or where a turbine valve (peripheral) or a positive displacement (reciprocating or rotary) type of pump is used, a suitable relief valve shall be installed on the pump discharge to prevent damage to the fire protection system.

A centrifugal-type pressure maintenance pump is preferable.

A fire pump shall not be used as a pressure maintenance pump.

Where emergency electric power is provided, the jockey pump motor should be connected to it.

A jockey pump is usually required with automatically controlled pumps.

32 Break Tanks

This chapter provides the minimum design guidelines for break tank design and installations. These guidelines are intended for use only at locations where a direct connection between a public water supply and a private fire protection system is prohibited.

Careful consideration should be given to providing a full- size suction tank or reservoir. Larger underground mains to fill the break tank, and additional equipment an appurtenance can result in a break tank installation cost approaching or exceeding that of a full-size suction tank. This, coupled with the decreased reliability of the fire protection system, may make the full-sized suction tank the more attractive alternative. Guidelines for the fire pump installation are contained in the Fire Pumps, section.

Break Tank

A break tank is an automatically filled tank which provides a suction water supply for a fire pump. The tank capacity is not adequate to supply the total sprinkler and hose stream demand for the required duration.

Break tanks provide a means of cross-connection control where jurisdictional authorities prohibit any direct connections between the public water supply and a private fire protection system, either for health or hydraulic reasons. This is accomplished by creating a physical break, or gap, between the public water supply and the private fire protection system. Water from the public water supply enters the break tank at a height above the tank overflow outlet "free falling" through the atmosphere into the tank. Since the water in the tank is considered potable, the top of the tank should be enclosed.

Since the water from the public supply is no longer under pressure once it enters the break tank, a fire pump taking suction from the break tank is needed to supply the fire protection water at adequate pressure. A booster fire pump, which previously took suction from the public water supply and was considered an adequate fire protection water supply, may not provide adequate water flow and pressure for fire protection if it is arranged to take suction from a break tank. The flow and pressure characteristics of the public water supply will no longer affect the actual flow and pressure characteristics of the fire protection water supply downstream from the fire pump.

Reliability of a break tank is less than that for an adequately sized suction tank because the automatic fill mechanisms are subject to mechanical failure. Also, the makeup water flowing into the break tank from the public supply can reduce pump performance due to aeration and turbulence problems.

Loss Experience

Stuck Fill Valve Causes Flooding

A 10,000 gal (37,850 dm^{-2}) break tank located in the basement of an office building supplied both domestic and fire protection water. The tank was filled automatically through a float valve connected to the public water supply through a 4 in. (150 mm) main. No overflow drain to a safe discharge or high water level alarm was provided. A guard making rounds discovered water in the basement, at which time the city curb box valve was closed. Carpeted areas and various pumps were damage.

Recommendations

General

1. Do not install a break tank to supply a fire pump which when operating at 150% capacity, exceeds the maximum capacity which could be supplied if the fire pump were connected directly to the public water supply.

Size of Tank

1. For new installations provide a separate break tank for each fire pump for exclusive use as a fire protection water contained between the level 2 ft. (0.6 m) above the pump suction pipe vortex plate and the level where the automatic fill valves begin to operate is sufficient to supply pump operation at 150% of the pump rated capacity for 15 minutes.
2. Tank design should allow a minimum of 15 ft. (4.6 m) horizontal distance between the centerline of fill outlets and the centerline of the pump suction pipe. This distance is needed to reduce the effect of turbulence and aeration at the pump suction pipe which could affect pump performance.

Fill Mechanism

1. Provide at least two automatic an easily accessible manual fill outlet to the tank. Size each fill outlet to independently supply makeup water at a rate not less than 150% of the pump rated capacity. Automatic fill valves should open when the water level reaches 6 in. (160 mm) below the full water line.
2. Automatic fill valves may be of the float, altitude, or other suitable design arranged for direct or pilot operation. Float valves of modulating type are preferred.

   There are two types of float valves; modulating and non-modulating. Modulating float valves operate using a pilot controlled system to balance influent and effluent rates.

   The float valve responds to changing water levels by draining small amounts of water from the pilot control system, which creates a pressure differential across the fill valve. The pressure differential determines valve position, and thus the flow rate into the tank.

It is essential that the fill valve not be oversized in order to maintain suitable pressure differentials for proper valve operation. Maximum recommended valved capacities can be obtained from valve manufacturers.

Non-modulating float valves open wide when the water level reaches the low float set point, and close fully when the water level reaches the low float set point, and close fully when the water reaches the high float set point. Because of this positive valve action, problems with water hammer may be introduced into the piping system feeding the fill valves.

For this reason, modulating type float valves are preferred over non-modulating type float valves.

3. Provide pipe wells for the float valves extending to within 6 in. (150 mm) of the tank bottom, sized at least 1 in. (25 mm) in diameter larger than the float to ensure proper float valve operation and to facilitate valve testing. When pipe wells are used, provide valved pipe well drains for each pipe well at least 4 in. (100 mm) in diameter located not more than 2 ft. (0.6 m) above the tank bottom arranged to discharge to a safe, frost-free location. For testing purposes and to make it impossible to close off pipe wells, cut four 1.5 in. (37 mm) holes above the pipe well drains. Opening the drain valves will allow testing of float valves by draining water from the pipe wells faster than it can be replenished through the four 1.5 in. (37 mm) holes in the pipe wells.
4. To reduce turbulence and aeration, design, and location of fill outlets should meet the following criteria.
   a. Pipe the fill outlets so that makeup water enters the tank perpendicular to the tank water surface.
   b. Locate fill outlets at the side of the tank opposite from the pump suction pipe, and a minimum of 15 ft. (4.6 m) horizontally from the pump suction pipe.
   c. Terminate the fill outlets at a level at least equal to two fill pipe diameters above the top of the overflow outlet, or at a height specified by the authority having jurisdiction, whichever is less.

      Turbulence and aeration will increase with increasing distance between the fill outlet and the tank water surface.
   d. Size the fill outlets so that water velocity at the maximum makeup water flow rate does not exceed 20 ft./sec (0.1 m/sec). Use Table f to determine the minimum fill pipe size. The fill line size, length, and friction loss in any backflow preventers or meters, together with the water supply filling the tank, will all be factors in refilling the tank.

Freeze Protection

5. Locate the tank, tank fill valves, fire pump, and associated piping in a heated area or otherwise protect them from damage if subject to freezing temperatures. A tank heater may be necessary. Ensure that water in float valve pipe wells, which may not be subject to convective heating if a tank heater is used, and pilot lines on automatic fill valves are also

adequately protected against damage from freezing. Data Sheet 3-2, "Water Tanks for Fire Protection," contains guidelines relating to freeze protection.

Fire Pump Suction Pipe Connection

1. Arrange pump suction piping for either horizontal shaft or vertical shaft turbine type pumps in accordance with Data Sheet provided by the pump manufacturer.
   For all horizontal shaft pumps, provide a 4 by 4 ft. (1.2 by 1.2 m) vortex plate at the lower end of the pump suction pipe elbow, located at a distance of one-half the suction pipe diameter, but no less than 6 in. (150 mm), above the bottom of the tank.

Additional Equipment

1. Provide a tank water level indicator visible to the fire pump operator.
2. Provide a low water level alarm in the tank, connected to a constantly attended location, to indicate when the tank water level drops to 9 in. (225 mm) below the full water line. This will signal the fact that more water is being pumped from the tank than is being replenished, allowing time to open the manual fill valve or to alert proper parties of this condition.
3. Baffling within the tank may be necessary if pump operation is unsatisfactory due to excessive turbulence or aeration.
4. Provide a tank overflow outlet capable of discharging to a safe, frost-free location an amount of water no less than that which can be supplied by the fill mechanism. Size the overflow drainpipe at least one diameter larger than the fill pipe size.
5. Provide a fire department pumper connection downstream of the break tank/fire pump installation.

Maintenance and Testing

1. Automatic fill valves should be tested every two weeks by opening the pipe well drain valve and flowing enough water until the automatic fill valve opens fully. Automatic fill valves may be tested alternately each week in conjunction with the usual weekly fire pump test.
2. Test and maintain the automatic fire pump installation in accordance with Data Sheet provided by the manufacturer.
3. Maintain the automatic fill valves in accordance with the manufacturer's recommendations.
4. Inspect, test and maintain all other fire protection system control valves and equipment in accordance with current guidelines.

33 Fire Service

This chapter covers the design, installation, arrangement, location, and preventive and corrective maintenance of private fire service mains and their appurtenances. This equipment includes manually operated valves, pressure regulating valves, hydrants, meters, and outside hose equipment including hose houses and cabinets. The hydraulics of underground mains, including hydrant and other orifice discharge coefficients and Hazen-Williams pipe coefficients, are covered in hydraulic calculations. Cross connections and the use of backflow preventers are covered by the local fire codes.

For the purpose of this chapter, a private fire service main is that pipe between a source of water and the water supply side of the first connection Aboveground to a sprinkler or standpipe system. When connected to a public water system, the private fire service main begins at a point designated by the public utility, usually at a manually operated valve near the property line. When connected to a fire pump, the main begins at the system side of the pump discharge valve. A main connected to a gravity or pressure tank begins at the system side of the tank's discharge valve.

Some mains are used to carry water for both fire service and industrial use. This type of main is called "combined service" mains.

Components

Meters

Some water utilities require the installation of flow meters in connections from public water mains to private fire service systems. When such meters are required, use Factory Mutual approved fire service type meters. Approved meters include both full registration meters and waterflow detector check valves. Each experiences friction loss which must be accounted for in the design of a private fire service system.

One approved full registration fire service meter uses a weighted check valve and a compound meter on a main line, combined with a single disc meter on a bypass line for the purpose of measuring small rates of flow. The weighted lever valve diverts small flows through the bypass meter for accurate measurement, and opens automatically to full capacity when there is a large demand for water. A compound type meter is available as an option on the bypass line.

With this type, small flows are measured accurately by a disc meter and the larger flows by a torrent (current type) meter.

Waterflow detector check valves are used where the water utility will accept a simple indication of high water flow, rather than a record of the actual amount of water used. Approved waterflow detector check valves generally operate as follows. The clapper starts to open when the differential pressure across the clapper assembly exceeds the established differential pressure of the check valve. Small flows that produce a differential pressure less than the established differential pressure will not move the clapper. These are directed through they bypass piping and are metered. Large flows, caused by water demand in the fire protection system, produce a greater differential pressure and move the clapper out of the waterway allowing the full flow to pass into the system unmetered.

Single Check Valves

Single check valves are used in fire protection piping systems to allow water to flow in one direction only. Gravity-operated swing check valves (see Figure 5) are kept open by water flow. Closure is effected by back pressure in the line or by the weight of the check mechanism when there is no flow.

Water authorities often require a single check valve or other backflow preventer at connections to a public water source.

This is to prevent the backflow of fire protection system water from nixing with the public water supply and the resultant possibility of contaminating the public supply.

Provide a single check valve in each connection to a public water source. This permits the fire pump or fire department pumper to raise the private fire protection system pressure to a higher level than that of the public mains.

Provide a single check valve at the discharge from fire pumps and at gravity tank and pressure tank connections. This will help to prevent water from the private fire protection main system from being lost through the tanks and pump supply.

Single check valves may be installed in either horizontal or vertical pipe unless restricted to a specific positioning as noted in the Approval Guide listing. For installation in vertical pipe the flow must be upward.

Install check valves so that they are accessible and, where underground, in watertight, frostproof concrete or masonry pits. Provide clearance around the main where it passes through the wall, and pack the annulus with a sealant. Provide manually operated valves, preferably of the indicating type, on both sides of the check valve to allow isolation for maintenance or repair. Frequently, there is a curb box valve at the connection to the public main which can serve as the upstream valve.

Provide ample clearance for valves with side mounted covers to permit clapper removal. Wafer check valves have a shorter body length and many do not have covers. Valves without covers must be removed from the system for clapper replacement or repair.

Antiwater-Hammer Check Valves

Water hammer is the term used to indicate the destructive forces, exemplified by pounding noises and vibration, which develop in a piping system where a column of non-compressible liquid flowing through a pipe line at a given pressure and velocity is stopped abruptly. When water hammer occurs, a high intensity pressure wave travels back through the piping system until it reaches a point of some relief, such as a larger diameter riser or piping main. The shock wave then surges back and forth between the point of relief and the point of impact until the destructive energy is dissipated in the piping system, sometimes in the form of broken piping. This violent action accounts for the piping noise and vibration. Water hammer can also occur without any audible sound.

Special antiwater-hammer check valves are approved for use in the discharge of a fire pump where conditions may cause severe water hammer with the usual single swing type check valve.

Generally, the antiwater-hammer check valve has a spring mechanism which automatically closes the valve disc at zero flow, before flow reversal occurs, thereby preventing surge and water hammer.

Because of the relatively high friction loss through antiwater-hammer check valves, consult the friction loss data tabulated in the Approval Guide and consider the loss in the design of the private fire service system. Some approved antiwater-hammer check valves are not suited for direct bolting to the body of butterfly valves because of clearance problems between the butterfly vane and the check valve clapper. Limit the use of antiwater-hammer check valves to installations on the discharge side of fire pumps or other water hammer problem areas.

For other situations use approved single swing check valves. The swing check valves have a lower friction loss, greater clearance between parts, and lower susceptibility to being obstructed.

Double Check Valves and Backflow Preventers Cross connections can result in the contamination of the public water source by the intermixing at the public source of private water supplies and the public water source. Cross connection control as dictated by water regulatory authorities can be achieved with double check valves or reduced pressure principle backflow preventers commonly referred to as RP devices.

Fire Department Pumper Connections

A fire department pumper connection, sometimes referred to as simply "pumper connection", is a connection through which a fire department can pump water into fire service mains or into a sprinkler system, thus raising the water pressure above that normally available.

Provide one or more pumper connections on private fire service systems. This is particularly important where the public water system is the only source of supply. To prevent loss of system pressure thorough the pumper connection, install an approved single check valve in each fire department connection, located as near as practicable to the point where it joins the system.

Do not provide a shutoff valve in the piping between the pumper connection and the fire service mains. Provide an approved automatic drip valve in the pipe between the check valve and the outside hose coupling of the pumper connection. Arrange the automatic drip system so that it discharges to a proper frost-free location.

Other information on the installation of fire department pumper connections and associated hose coupling threads is covered in Installation Sprinkler Systems.

Manually Operated Valves

Manually operated valves are used to control individual sources of water supply, isolate equipment such as check valves for maintenance, and allow sectionalizing of water main systems in the event of a break or for the purpose of making repairs or extensions.

Gate Valves and Butterfly Valves. Gate valves and butterfly valves with associated indicating mechanisms are commonly used in fire service mains. With the gate valve in its fully open position, the disc is lifted completely out of the waterway. Butterfly valves utilize a quarter turn disc to control flow. With the butterfly valve in its open position, the disc sits on the waterway, parallel to the direction of flow.

A Factory Mutual approved Post Indicator Valve Assembly (PIVA) is available only as a factory-assembled unit with true open-shut orientation. The assembly consists of a buried, quarter turn valve and an above grade indicator operator. If excessive force is applied to the handle, one of the series of successively weaker shear joints incorporated in the drive train will break so as to preserve the correct orientation between indicator and valve position. Parts are so designed that they can be assembled only one way, always to give correct indication. In case the post is broken off above grade, fail-safe features will keep the valve intact and a spring will hold an open valve open.

Indicator Posts. Indicator posts are controls, extending Aboveground, for operating underground fire service main valves. As an important feature of their design, a target or indicator, visible through an opening in the post, indicates whether the valve is open or shut.

The indicator post consists of a cast iron barrel, an extension flange which allows the height of the barrel to be regulated, and a wrought iron or mild steel stem which operates the underground valve and simultaneously moves the target. Indicator posts are available for installation on existing gate valves.

Installing Manually Operated Valves. For underground fire service main use, install either a post indicator valve assembly or a gate valve with approved indicator post to readily show the open or closed position of the valve. If a valve must be placed where an indicator post cannot be used, such as in a road, use an indicating valve such as an approved Indicating Butterfly Valve (IBV) or an Outside Screw and Yoke Gate Valve (OS&Y) installed in a watertight, frostproof concrete or masonry pit.

Indicating valves are preferred. Therefore, limit the use of non-indicating type underground gate valves to branch mains to hydrants and building lead-ins. Such valves are normally non-rising stem type and require a key wrench for operation.

A curb box with cover plate at grade level provides access for the wrench to the valve stem nut.

All approved PIVA, IBV, and OS&Y valves open counterclockwise. Currently approved Non-Rising Stem (NRS) gate valves for operation with key wrenches through curb boxes or with indicator posts are manufactured to open counterclockwise. In both new fire service main system installations and valve replacement situations, install approved valves which open in a counterclockwise direction.

In existing installations where all valves open clockwise or a mixture of opening directions are present, clearly identify the direction of opening. A sign securely attached to the indicator post which shows the direction of opening is an acceptable form of identification. For curb box installations, the cover plate can be appropriately marked or a sign installed on a nearby post of building wall.

Tapping Valves. The use of a tapping valve allows a new connection to be made to an existing water main without shutting off the water.

Pressure Regulating Valves

Pressure regulating valves are automatic devices which control the downstream pressure in a pipeline supplied from a higher pressure source. They open and close automatically according to flow demands. These valves are often installed in public water distribution systems to control the flow of water from high to low pressure zones or balance the pressures from standpipes and reservoirs of different elevations.

Where direct connections to high pressure mains might impose excessive pressure on sprinklers and yard piping, use pressure regulating valves in fire service mains.

Pressure regulating valves are of two types: pilot operated fixed pressure and differential.

For fire protection service, pilot operated fixed pressure valves are recommended to take full advantage of the water supply available at the regulated pressure.

The Pilot Operated Fixed Pressure type valve maintains a constant outlet pressure at any rate of flow within the capacity of the valve. Pressure control is provided by a pilot device that may or may not be separate from the body of the main valve. Pressure reduction depends on friction loss created by flow through the main valve openings. The effective area varies with the position of the valve piston. If the rate of flow should increase sufficiently to drop the inlet pressure to the set point (maximum desired on low pressure side) or below, regulation will cease and the valve will open wide. Under these conditions, outlet pressure will equal inlet pressure minus the friction loss through the wide open valve, which is usually assumed to be equal to that of a globe valve of the same size.

The outlet or delivery pressure is set by adjustment of screw, which increases or decreases the tension of a spring.

Throttle valve is used to throttle high pressure port to suit operating conditions and control the regulating valve's speed of operation.

Whenever the outlet pressure, which is transmitted through port to diaphragm, increases sufficiently to overcome the fore exerted by spring, valve rises to its seat and prevents water flow through port. Inlet pressure from port through throttle valve will then cause piston to move toward the closed position. If the outlet pressure continues to keep valve on its seat, piston will move until the flow stops completely.

A reduction in outlet pressure below that fixed by adjusting screw allows spring to expand, opening valve the higher pressure above piston (through port from the valve inlet) then escapes through port, valve, and port. With pressure above piston reduced, the inlet pressure below piston causes it to move upward and permit the flow of water. With steady flow, piston remains in a floating position with openings just adequate to maintain the downstream pressure.

The Differential Valve maintains a constant pressure drop between the inlet and outlet for all rates of flow and may be either direct acting (valve piston attached to a diaphragm) or pilot operated.

Pressure regulating valves of several manufacturers differ considerably physically, but the basic principles of operation are similar. Various methods are employed to promote smooth operation and reduce valve chatter. Pressure regulation is usually possible within 2 psi (14kPa, 0.14 bar) of the set point.

When a pressure regulating valve is to be installed in a private fire service connection, determine, by testing, the characteristics of the inlet supply and carefully estimate the maximum demands for fire protection and industrial use. The set point is fixed by two limits: (a) the maximum pressure to be permitted on the yard piping, and/or (b) the residual pressure expected on the outlet side of the valve at the desired rate of flow with the valve wide open. Provide these data to the manufacturer, with information on the static and residual pressure at the maximum rate of demand on the inlet side of the valve. This will enable the manufacturer to provide a valve of adequate capacity and suitable characteristics for the specific conditions.

It may be necessary to install a small pressure regulating valve (approximately 2 in. [51 mm]) in parallel with a larger valve to achieve pressure regulation at small flows. For example, one manufacturer's 8 in. (203 mm) valve requires a minimum 75 gal/min (284 dm^3/min) flow for operation.

Thoroughly flush the underground main to remove all accumulations of scale or dirt before installing. If service is taken from a pipeline carrying raw or unscreened water, provide an approved strainer. Otherwise, dirt may enter the regulating valve and score the internal parts or plug small openings. For severe conditions, connect two strainers in parallel.

Equip connections from high pressure mains with readily accessible control valves.

Locate meters on the yard side of the pressure regulating equipment.

Locate pressure regulating valves in dry accessible pits, arranged for convenient testing and servicing. Provide a test drain, of minimum size one half the pressure regulating valve size, on the downstream side of the valve. An extension pipe, the same size as the test drain, with hose fittings may be needed to discharge the water from the test drain.

Provide pressure gauges on both sides of the valve, trimmed in a manner similar to the test gauges on double check valve installations. If the regulating valve is actuated by a pilot valve having an external control pipe, provide a pressure gauge on the control pipe close to the pilot so that the pressure acting on the diaphragm will be indicated. The gauge on the inlet side is necessary to determine the accuracy of the regulation at various rates of flow.

Provide a bypass around the regulating valve. The valve in the bypass would normally be shut but could be opened in an emergency.

Provide relief valves near the regulating valve pits and at or near the extremities of large systems to protect yard and sprinkler piping from excessive pressure. Arrange the discharge piping from the relief valves so that the discharge will be visible. Determine the capacity and location of relief valves by careful study. Ordinarily, a capacity of one-fourth of the recommended maximum flow of the pressure regulating valve should be satisfactory.

Shows a typical layout with one valve. With this arrangement, the fluctuations in industrial demand can be expected to cause frequent movement of the valve mechanism, helping to maintain the equipment in good condition.

The periodic waterflow alarm tests recommended in Firesafety Inspections and Maintenance of Sprinkler Systems also help to exercise the regulating valve. This is in addition to regular testing through the test drain.

Hydrants

Hydrants are used to supply water for manual firefighting. Outdoor hazards for which hose stream protection is especially needed include storage of combustible materials and equipment, freight cars and loading platforms and Aboveground flammable liquid tanks. Hose streams from hydrants or hose connections on the roof are needed to protect combustible roof structures and equipment such as cooling towers, oil-filled transformer installations, and large areas of smooth-surfaced roof coverings. For unsprinklered buildings, hose streams are usually the main defense against fire.

Dry barrel hydrants are used in areas where freezing temperatures are experienced. Water is admitted into the barrel only when the hydrant valve is opened to prevent the water which remains in the barrel from freezing after the hydrant is closed, a small drain hole opens as the hydrant valve is closed. "Traffic" model hydrants are available incorporating sections of intentional weakness at ground level so that if struck by a vehicle, the hydrant will break off aboveground level without the hydrant valve opening.

The most common kind of dry barrel hydrant is the compression type.

Wet barrel hydrants are used in areas where the temperature remains above freezing. They usually have a compression valve at each outlet, but may have one such valve in the bonnet that controls the flow of water to all outlets.

Approved hydrants have either two or three hose outlets. Provide hydrants which have three outlets with approved hydrant valves. Do not install hydrants which have pumper-suction outlets in private fire protection systems. The draft created by pumpers connected to pumper-suction outlets of a hydrant could deplete the water supply to sprinklers.

34 Fire Protection for Storage of Rubber Tires

Rubber is one of the most combustible substance and special care must be given where is stored.

In buildings with storage exceeding 15 ft. (4.6 m) through 20 ft. (6 m) in height. The fire protection design should include one-hour fireproofing or one sidewall sprinkler head directed to one side of the column at 15 ft. (4.6 m) level.

In buildings with storage exceeding 20 ft. (6 m) in height.

Two hour fireproofing for the entire length of the column, and including connections with other structural members; on two sidewall sprinkler heads, one at the top of the column and the other at the 15-ft. (4.6 m) level, both directed to the side of the column.

In buildings where storage in fixed racks is protected by in-rack sprinklers. The protection of columns is not required.

In fully sprinklered buildings stored tires shall be segregated from other combustible storage by aisles at least 8 ft. (2.4 m) wide. In the absence of sprinkler systems, stored tires shall be cut off by fire walls.

When tires are stored up to 15 ft. (4.6 m) high, walls between adjacent warehouse areas and between manufacturing and warehouse areas shall have not less than a 4-hr. fire rating. When tires are stored over 15 ft. (4.6 m) high, walls between manufacturing and warehouse areas shall have a fire rating of not less than 6 hrs.

Pillars shall be not more than 50 ft. (15 n) in width except that piles along a wall shall not be more than 25 ft. (7.6 n) in width.

The width of main aisles between piles shall be not less than 8 ft. (2.4 m).

The clearance from the top of storage to sprinkler deflectors shall be not less than 3 ft. (.9 m).

Storage clearance in all directions from roof structures shall be not less than 3 ft. (.9 m).

Storage clearance from units heaters, radiant space heaters duct furnaces and flues shall not be less than 3 ft. (.9 m) in all directions, or shall be in accordance with the recommendation by the authorities having jurisdictions.

Clearance shall be maintained to lights or light fixtures to prevent possible ignition.

Not less than 24 in. (.6 m) clearance shall be maintained around the path of fire door travel unless a hassifide is provided.

Automatic sprinklers shall be installed in accordance with NFPA 13, except as modified by local codes.

Sprinkler discharge densities and areas of application shall be in accordance with NFPA No.13 or recommended by the authorities having jurisdictions.

In buildings which are occupied in part for tire storage, where only a portion of the sprinkler system is hydraulically designed, the design area shall extend not less than 15 ft. (4.6 m) beyond the area occupied by the tires.

The area protected by a single system of sprinklers in racks (in-rack sprinklers) shall not exceed 40,000 sq ft. (3716 m^{-2}) of floor area occupied by the racks, including aisles, regardless of the number of intermediate sprinkler levels.

When sprinklers are installed in racks, separate indicating gate valves and drains shall be provided for ceiling sprinklers and sprinklers in racks, except such drains and valves are not required for small in-rack installations of less than 20 sprinklers.

Water demand of sprinklers installed in racks shall be added to ceiling sprinkler water demand at the point of connection.

Sprinklers in racks shall be ordinary temperature classification with nominal½ in. (12.7 mm) orifice size pendent or upright.

The number of sprinklers and the pipe sizing on a line of sprinklers in racks is restricted only by hydraulic calculations, and not by any piping schedule.

Water shields shall be provided directly above in-rack sprinklers, or listed sprinklers equipped with water shields shall be sued when there is more than one level of in-rack sprinkler. In-rack sprinkler deflectors shall be located at the same level as the bottom of the pallet support to maintain an unobstructed clear space of at least 4 in. (102 mm). In-rack sprinklers shall be located at least 2 ft. (0.6 mm) from rack uprights.

In-rack sprinklers at one level only for storage up to and including 20 ft. (6 mm) high shall be located at one half to two thirds of the storage height.

Maximum horizontal spacing of sprinklers in racks shall be 8 ft. (2.4 m).

Sprinklers in racks shall discharge at not less than 30 psi (207 kPa) for all classes of commodity.

Water demand for sprinklers installed in racks shall be based on simultaneous operation of the most hydraulically remote 12 sprinklers when only one level is installed in racks.

High expansion foam systems may be installed in addition to automatic sprinklers. In that case, a reduction in sprinkler discharge density will be allowed. High expansion foam systems shall be automatic in operation.

Detectors shall be installed at the ceiling at the recommended spacing or in accordance with NFPA 72E.

Detection systems, concentrate pumps, generators and other system components essential to the operation of the system shall have an approved standby power source.

The rate of water supply shall be sufficient to provide the required sprinkler discharge density over the required area of application plus provision for generation of high expansion foam and in-rack sprinklers.

Total water supplies shall include provision for not less than 750 gal/min (2835 L) for hose streams, in addition to that required for automatic sprinklers and foam systems.

Water supplies shall be capable of supplying the demand for sprinkler systems and hose streams for not less than 3 hrs.

Where dry pipe systems are used, the area of sprinkler application shall be increased by not less than 30 percent.

Where automatic sprinkler protection is provided, snail hose (1½ in.) (38 mm) shall be provided to reach any portion of the storage area. Small hose may be supplied from:

1. Hydrants
2. A separate piping system for snail hose stations
3. Valved hose connections on sprinkler risers where such connections are made upstream of sprinkler control valves
4. Adjacent sprinkler systems.

In locations where snail hose is provided, portable fire extinguishers for Class A fires may be omitted in storage areas.

At locations without public hydrants, or where hydrants are not within 250 ft. (76 m), private hydrants shall be installed.

Automatic sprinkler systems and foam systems where provided shall have approved central station, auxiliary, remote station or proprietary waterflow alarm service.

Alarm service shall comply with one of the following: NFPA 71, 72A, 72C or 72D.

Arrangements shall be made to permit rapid entry into the premises by the municipal fire department, police department, or other authorized personnel in case of fire or other emergency. Plant emergency organizations where provided shall be instructed and trained in the following procedures:

1. Maintaining the security of the premises.
2. Means of summoning outside aid immediately in an emergency.
3. Use of portable extinguishers and snail hose lines or snail fires and pop-up operations.
4. Operation of the sprinkler system and water supply equipment.
5. Use of material handling equipment while sprinklers are still operating to effect final extinguishment.
6. Supervision of sprinkler valves after the system is turned off so that the system can be reactivated if rekindling occurs.
7. A fire watch shall be maintained when the sprinkler system is not in service.

Power-operated industrial trucks shall comply with NFPA 505, Including Type Designations and Areas of Use.

Wood pallets shall be stored in accordance with the requirements of NFPA 231.

When welding or cutting operations are necessary, the necessary precautions shall be followed. When possible, work shall be removed to a safe area.

Welding, soldering, brazing, and cutting may be performed on rack or building components which cannot be removed, provided no storage is located below and within 25 ft. (7.6 m) of the working area, and flameproof tarpaulins enclose this section. During any of these operations the sprinkler system shall be in service. Extinguishers suitable for Class A fires with a minimum rating of 2A and charged inside hose lines where provided shall be located in the working area. A fire watch shall be maintained during their operations and for not less than 30 min. following completion of open flame operation.

Rubbish, trash, and other waste material shall be disposed of at regular intervals.

Smoking shall be strictly prohibited, except in locations prominently designated as smoking areas. "No Smoking" signs shall be posted in prohibited areas.

Fire walls, fire doors, and floors shall be maintained in good repair at all times.

The sprinkler system and the water supplies shall be maintained and serviced.

33 Fire protection Signaling Systems

Protective signaling systems have been in use for many years, and their value, in connection with the fire hazard, is recognized as applying both to the protection of life and to the prevention of property damage.

The general experience with all protective signaling systems is that their dependability is directly proportional to the amount and effective character of maintenance, testing, and supervision, and that these functions are best performed when the owner or occupant of the protected premises contracts for service with an outside agency known to be qualified in the handling of central station signaling service.

The central station shall be equipped with the necessary instruments for automatically receiving and recording all signals. The time of receipt of signals shall also be recorded automatically or manually. Circuit adjusting means for emergency operating may either be automatic or be provided through manual operation upon receipt of a trouble signal.

Two independent means shall be provided to retransmit a fire alarm signal received from a protected premises to the appropriate public fire service communication center.

Alarm signals initiated by manual fire alarm boxes, automatic fire detectors, waterflow from the automatic sprinkler system or actuation of other fire extinguishing systems or equipment shall be treated as fire alarms.

The central station should:

1. Immediately retransmit the alarm signal to the fire department.
2. Promptly dispatch a runner or technician (travel time not to exceed one hour) to the protected premises, when equipment must be manually reset.
3. Notify the subscriber by the quickest available method.
4. Provide written notice to the subscriber and the authority having jurisdiction.
5. Notify such other parties as the authority having jurisdiction may require.

Upon failure to receive a guard's regular signal within a 15 minute maximum grace period, the central station should:

1. Communicate at once with the protected premises by telephone, or by calling back on the system circuit.

2. If communications cannot be promptly established, immediately dispatch a runner (travel time not to exceed half an hour) to investigate the delinquency.
3. Report all delinquencies in writing to the subscriber and the authority having jurisdiction.

Failure of the guard to follow a prescribed route in transmitting signals shall be handled as a delinquency.

Upon receipt of a supervisory signal from a sprinkler system, other extinguishing systems or other equipment, the central station should:

1. Communicate immediately with person(s) designated by the subscriber.
2. Dispatch a runner or maintenance person (arrival time not to exceed one hour) to investigate unless abnormal condition is restored to normal in accordance with scheduled procedure determined by (a) above.
3. Notify the fire department if required.
4. Notify the authority having jurisdiction as to the nature of the signal, time of occurrence, and restoration of service when equipment has been out of service for eight hours or more.

Upon receipt of trouble signals or other signals pertaining solely to matters of equipment maintenance of the signaling systems, the central station should:

1. Investigate immediately by telephone, or if necessary,
2. Dispatch personnel to arrive within four hours to initiate maintenance.
3. Provide written notice to the subscriber and the authority having jurisdiction as to the nature of the interruption, time of occurrence and restoration of service, when the interruption is more than eight hours.

Automatic fire alarm systems shall be arranged to transmit a signal designating the building protected, and shall also indicate, either in the signal or by approved visual means in or at the building, the floor, section or subdivision from which the signal originated, except that the authority having jurisdiction may, upon application, waive this specific indication where the area, height or special conditions of occupancy are such that this indication is not deemed to be an essential feature of alarm protection of the property.

36 Heat Detectors

Fire is a phenomenon which occurs when a substance upon reaching a critical temperature reacts chemically, as for example, with oxygen, to produce heat, flame, light, smoke, water vapor, carbon monoxide, carbon dioxide, or other products and effects.

An automatic fire detector is a device designed to detect the presence of fire and initiate action.

Typical examples of fixed temperature sensing elements are:

1. Bimetallic. A sensing element comprised of two metals having different coefficients of thermal expansion arranged so that the effect will be deflection in one direction when heated and in the opposite direction when cooled.
2. Electrica1 Conductivity. A line-type or spot-type sensing element whose resistance varies as a function of temperature.
3. Fusible Alloy. A sensing element of a special composition metal which melts rapidly at the rated temperature.
4. Heat-Sensitive Cable. A line-type device whose sensing element comprises, in one type, two current-carrying wires held separated by a heat-sensitive insulation which soften at the rated temperature, thus allowing the wires to make electrical contact. In another type a single wire is centered in a metallic tube and the intervening space filed with a substance which, at a critical temperature, becomes conductive, thus establishing electrical contact between tube and the wire.
5. Liquid Expansion. A sensing element comprising a liquid marked expansion in volume in response to temperature increase.

A rate compensation detector is a device which will respond when the temperature of the air surrounding the device reaches a predetermined level, regardless of the rate of temperature rise.

A typical example is a spot-type detector with a tubular casing of a metal that tends to expand lengthwise as it is heated, and an associated contact mechanism that will close at a certain point in the elongation. A second metallic element inside the tube exerts an opposing force on the contacts, tending to hold them open. The forces are balanced in such a way that on a slow rate of temperature rise, there is more time for heat to penetrate to the inner element, which therefore inhibits contact closure until the total device has been heated to its rated temperature level. However, on a fast rate of temperature rise, there is not as much time for heat to penetrate to the inner element, which therefore exerts less of an inhibiting effect so that contact closure is obtained when the total device has been heated to a lower level. This, in effect, compensates for thermal lag.

A rate-of-rise detector is a device which will respond where the temperature rises at a rate exceeding a predetermined amount.

Typical examples are:

1. Pneumatic Rate-of-Rise Tubing. A line-type detector comprising small diameter tubing, usually copper, which is installed on the ceiling or high on the walls through the detected area. The tubing is terminated in a detector unit containing diaphragms and associated contacts set to actuate at a predetermined pressure. The system is sealed except for calibrated vents which compensate for normal changes in temperature.
2. Spot-type Pneumatic Rate-of-Rise Detector. Consisting of an air chamber, diaphragm, contacts, and compensating vent in a single enclosure.
3. Thermoelectric Effect Detector. A device whose sensing element comprises a thermocouple or thermopile unit which produces an increase in electric potential in response to an increase in temperature. This potential is monitored by associated control equipment, and an alarm is initiated when the potential increases at an abnormal rate.

Temperature Classification

Heat detectors of the fixed-temperature or rate-compensated spot-pattern type shall be classified as to the temperature of operation and marked with the appropriate color code.

Where the overall color of a detector is the same as the color code marking required for that detector, either one of the following arrangements, applied in a contrasting color and visible after installation, shall be employed:

1. A ring on the surface of the detector.
2. The temperature rating in numerals at least 3/8 in. (9.5 mm) high.

Spot-type heat detectors shall be located upon the ceiling not less than 4 in. (100 mm) from the side wall or on the side walls between 4 in. (100 mm) and 12 in. (300 mm) from the ceiling. (See Figure A-3-4.1 in Appendix A.)

Exception No. 1: In the case of solid joist construction, detectors shall be mounted at the bottom of the joists.

Exception No. 2: In the case of beam construction where beams are less than 12 in. (300 mm) in depth and less than 8 ft. (2.4 m) on center, detectors may be installed on the bottom of beams.

Line-type heat detectors shall be located upon the ceiling or on the side walls not more than 20 in. (500 ram) from the ceiling.

High Temperature Areas

Detectors having fixed temperature or rate compensated elements shall be selected in accordance with Table 3-3.1 for the maximum ceiling temperature that can be expected.

Smooth Ceiling Spacing. One of the following rules shall apply:

1. The distance between detectors shall not exceed their listed spacing and there shall be detectors within a distance of one-half the listed spacing, measured at a right angle, from all walls or partitions extending to within 18 in. (460 mm) of the ceiling; or
2. All points on the ceiling shall have a detector within a distance equal to 0.7 times the listed spacing. This will be useful in calculating locations in corridors or irregular areas.

Irregular Areas. For irregular shaped areas the spacing between detectors may be greater than the listed spacing, provided the maximum spacing from a detector to the furthest point of a side wall or corner within its zone of protection is not greater than 0.7 times the listed spacing.

On ceilings 10 ft. (3 m) to 30 ft. (9.1 m) high, heat detector spacing shall be reduced in accordance with charts provided by the manufacturer or recommended by the authorities having jurisdictions. Except for the following detectors which rely on the integration effect:

1. Line-type electrical conductivity detectors.
2. Pneumatic rate-of-rise tubing.
3. Series connected thermoelectric effect detectors.

In these cases, the manufacturer's recommendations shall be followed for appropriate alarm point and spacing.

Solid Joist Construction. The spacing of heat detectors, when measure at right angles to the solid joists, shall not exceed 50 percent of the smooth ceiling spacing allowable.

If the beams project more than 18 in. (460 mm) below the ceiling, and are more than 8 ft. (2.4 m) on centers, each bay formed by the beams shall be treated as a separate area.

Sloped Ceilings. A row of detectors shall first be spaced and located at or within 3 ft. (0.9 m) of the peak of the ceiling, measured horizontally. The number and spacing of additional detectors, if any, shall be based on the horizontal projection of the ceiling in accordance with the type of ceiling construction.

The shed shall have a row of detectors located on the ceiling within 3 ft. (0.9 m) of the high side of the ceiling measured horizontally, spaced in accordance with the type of construction. Remaining detectors, if any, shall then be located in the remaining area on the basis of the horizontal projection of the ceiling.

Ionization Smoke Detection Principle. Smoke detectors utilizing the ionization principle are usually of the spot type. As ionization smoke detector has a small amount of radioactive material which ionizes the air in the sensing chamber, thus rendering it conductive and permitting a current flow through the air between two charged electrodes. This gives the sensing chamber an effective electrical conductance. When smoke particles enter the ionisation area, they decrease the conductance of the air by attaching themselves to the ions, causing a reduction in mobility. When the conductance is less than a predetermined level, the detector responds.

Photoelectric Light Obscuration Smoke Detection Principle. Smoke detectors utilizing the photoelectric light obscuration principle consist of a light source which is projected onto a photosensitive device. Smoke particles between the light source and the photosensitive device reduce the light reaching the device, causing the detector to respond.

Projected Beam Smoke Detector. This is a line-type obscuration smoke detector, where the light beam is projected across the area to respond.

Photoelectric Light Scattering Smoke Detection Principle. Smoke detectors utilizing the photoelectric light scattering principle are usually of the spot type. They contain a light source and a photosensitive device so arranged that the light rays do not normally fall onto the photosensitive device.

When smoke particles enter the light path, light strikes the particles and is scattered onto the photosensitive device, causing the detector to respond.

Resistance Bridge Smoke Detection Principle. A smoke detector utilizing the resistance bridge principle is usually of the spot type. It responds to a combination of smoke particles and moisture which falls onto an electrical bridge grid. As conductive substances fall onto the grid, they increase the conductance of the grid and cause the detector to respond.

Cloud Chamber Smoke Detection Principle. A smoke detector utilizing the cloud chamber principle is usually of the sampling type. An air pump draws a sample of air from the protected areas into a high humidity chamber within the detector. After the air is in the humidity chamber; the pressure is lowered slightly. If smoke particles are present, the moisture in the air condenses on them forming a cloud in the chamber. The density of this cloud is then measured by a photoelectric principle. When the density is greater than a predetermined level, the detector responds.

Location and Spacing. The location and spacing of smoke detectors shall result from an evaluation based on engineering judgment supplemented by the guidelines detailed in this standard. Ceiling shape and surfaces, ceiling height, configuration of contents, burning characteristics of combustible material present, and ventilation are some of the conditions that shall be considered.

Where the intent is to protect from a specific hazard, the detectors may be installed closer to the hazard in a position where the detector will readily intercept the smoke.

The possible effect of smoke stratification at levels below the ceiling shall also be considered.

Spot-type smoke detectors shall be located on the ceiling not less than 4 in. (100 mm) from a sidewall to the near edge, or if on a sidewall, between 4 in. and 12 in. (100 mm and 300 mm) down from the ceiling to the top of the detector. Except in the case of solid joist construction, detectors shall be mounted at the bottom of the joists and in the case of beam construction where beams are less than 12 in. (300 mm) in depth and less than 8 ft. (2.4 m) on centre, detectors may be installed on the bottom of beams.

Projected beam-type smoke detectors shall be located with their projected beams parallel to the ceiling, and not more than 20 in. (500 mm) below the ceiling. Except when a beams is installed vertically, or at any angle needed to afford protection of the hazard involved. Such as a vertical beams through the open shaft area of stair well where there is a clear vertical space inside the handrails.

The beam length shall not exceed the maximum permitted by the equipment listing.

Where mirrors are used with projected beams, the total listed length of the beams shall be progressively reduced by 33⅓ percent for each mirror used.

Beams shall be adjustable for total light cut-off values ranging from 5 percent on up to at least 40 percent.

Settings of½ percent per ft. (0.3 m), or less (more sensitive), shall be used whenever conditions permit.

Not more than 60 ft. (18.3 m) of beam length shall be considered in determining percent cut-off per ft. While the cut-off is not linear since each increment of beam length affects the next increment, linear calculations may be used for purposes of this standard.

Each sampling port of a sampling-type smoke detector shall be treated as a spot-type detector for the purpose of location and spacing.

Smooth Ceiling Spacing

Spot-type Detectors. On smooth ceilings, spacing of 30 ft. (9.1 m) may be used as a guide. In all cases, the manufacturer's recommendations shall be followed. Other spacing may be used depending on ceiling height, different conditions or response requirements.

Projected Beam-type Detectors. For location and spacing of projected beam-type detectors, the manufacturer's installation instructions shall be followed.

Solid Joist Construction

Ceiling construction where joists are 8 in. (200 mm) or less in depth shall be considered equivalent to a smooth ceiling.

If joists exceed 8 in. (200 mm) in depth, the spacing of spot-type detectors in the direction perpendicular to the joists shall be reduced. If the projected light beams of line-type detectors run perpendicular to the joists, no spacing reduction is necessary; however, if the projected light beams are parallel to the joists, the spacing between light beams shall be reduced.

Ceiling construction where beams are 8 in. (200 mm) or less in depth shall be considered equivalent to a smooth ceiling.

If beams are over 8 in. (200 mm) in depth, the spacing of spot-type detectors in the direction perpendicular to the beams shall be reduced. The spacing of line-type detectors run perpendicularly to the beams need not be reduced; however, if the line-type detectors are run parallel to the beams the spacing shall be reduced.

If the beams exceed 18 in. (460 mm) in depth and are more than 8 ft. (2.4 m) on centers, each bay shall be treated as a separate area requiring at least one spot-type or line-type detector.

Sloped Ceilings. Detectors shall first be spaced and located within 3 ft. (0.9 m) of the peak, measured horizontally. The number and spacing of additional detectors, if any, shall be based on the horizontal projection of the ceiling.

Detectors shall first be spaced an located within 3 ft. (0.9 m) of the high side of the ceiling, measured horizontally. The number and spacing of additional detectors, if any, shall be based on the horizontal projection of the ceiling.

Flame detectors shall be installed in all areas where they are required either by the appropriate NFPA standard, or by authority having jurisdiction.

Flame Detector. A device which responds to the appearance or radiant energy visible to the human eye (approximately 4000 to 7700 Angstroms) or to radiant energy outside the range of human vision.

Flame Flicker Detector. A photoelectric flame detector including means to prevent response to visible light unless the observed light is modulated at a frequency characteristic of the flicker of a flame.

Infrared Detector. A device whose sensing element is responsive to radiant energy outside the range of human vision (above approximately 7700 Angstroms).

Photoelectric Flame Detector. A device whose sensing element is a photocell which either changes its electrical conductivity or produces an electrical potential when exposed to radiant energy.

Ultraviolet Detector. A device whose sensing element is responsive to radiant energy outside the range of human vision (below approximately 4000 Angstroms).

Flame detectors are sensitive to glowing embers, coals, or actual flames, which radiate to the detectors energy of sufficient intensity and spectral quality to initiate action.

The detectors shall respond to the radiation from the area of fire that is to be detected. This usually involves expert application engineering. The time in which a fire must be detected and the area or intensity thereof may have to be related to the capabilities of associated extinguishing media and equipment.

Except as otherwise permitted herein, flame detectors shall not be spaced beyond their listed or approved maximums.

Closer spacing shall be utilised where the structural and other characteristics of the protected hazard would otherwise impair the effectiveness of the detection.

Flame detectors shall be so designed and installed that their field of vision will be sufficient to assure detection of a specified area of fire.

Where conveyance of materials on chutes or belts, or in ducts or tubes, or otherwise, to or past a detector is involved, spacing considerations will not govern, but strategic placement of detectors is required to assure adequate detection.

Since flame detectors are essentially line-of-sight devices, special care shall be taken in applying them to assure that their ability to respond to the required area of fire in the zone which is to be protected will not be unduly compromised by the presence of intervening structural members or other opaque objects or materials.

The overall situation shall be reviewed frequently to assure that changes in structural or usage conditions that could interfere with fire detection capabilities are remedied promptly.

Flame detectors shall have such spectral and optical response capabilities that they will initiate action from the specific spectral emission which occurs when the particular fuel(s) of the protected hazard is afire.

Detectors shall be designed, protected, or serviced so that interference with reception of radiation will not occur so as to prevent operation.

Where necessary, detectors shall be shielded or otherwise arranged to prevent action from unwanted radiant energy.

When used in outdoor applications, detectors shall be shielded in a fashion to prevent diminishing sensitivity by rain, snow, etc., and yet allow a clear field of vision of the hazard area.

Fire-gas detectors of the semiconductor type respond to either oxidizing or reducing gases by creating electrical changes in the semiconductor. The subsequent conductivity change of the semiconductor causes actuation.

Fire-gas detectors of the catalytic element type contain a material which in itself remains unchanged by accelerates the oxidation of combustible gases. The resulting temperature rise of the element causes actuation.

The location and spacing of fire-gas detectors shall result from an evaluation based on engineering judgment supplemented by the guidelines detailed in this standard. Ceiling shape and surfaces, ceiling height, configuration of contents, burning characteristics of combustible material present, and ventilation are some of the conditions that shall be considered.

Where the intent is to protect from a specific hazard, the detector(s) may be installed closer to the hazard in a position where the detector will readily intercept the fire gases.

The possible effect of gas stratification at levels below the ceiling shall also be considered.

Spot-type fire-gas detectors shall be located on the ceiling not less than 4 in. (100 mm) from a sidewall to the near edge, or if on a sidewall, between 4 in. and 12 in. (100 mm and 300 mm) down from the ceiling to the top of the detector. Except in the case of solid joist construction, detectors shall be mounted at the bottom of the joists. And in the case of beam construction where beams are less than 12 in. (300 mm) in depth and less than 8 ft. (2.4 m) on center, detectors may be installed on the bottom of beams.

Each sampling port of a sampling-type fire-gas detector shall be treated as a spot-type detector for the purpose of location and spacing.

Smooth Ceiling Spacing and Spot-type Detectors. On smooth ceilings, spacing of 30 ft. (9.1 m) may be used as a guide. In all cases, the manufacturer's recommendations shall be followed. Other spacing may be used depending on ceiling height, different conditions or response requirements.

Ceiling construction where joists are 8 in. (200 mm) or less in depth shall be considered equivalent to a smooth ceiling.

If joists exceed 8 in. (200 mm) in depth, the spacing of spot-type detectors in the direction perpendicular to the joists shall be reduced.

Ceiling construction where beams are 8 in. (200 mm) or less in depth shall be considered equivalent to a smooth ceiling.

If beams are over 8 in. (200 mm) in depth, the spacing of spot-type detectors in the direction perpendicular to the beams shall be reduced.

If the beams exceed 18 in. (460 mm) in depth and are more than 8 ft. (2.4 m) on centers, each bay shall be treated as a separate area requiring at least one spot-type detector.

37 Fire Protection for Data Processing Centers

Data processing equipment has become a vital and commonplace tool for business, industry, government, and research in groups in recent years. The use of such equipment is a direct result of the technological breakthroughs which have made the equipment available and the increased complexity of modern business, industrial, governmental and research needs.

Particularly pertinent are the increasing number of variables which must be taken into consideration in everyday decisions—overlooking any one item may spell the difference between profit and loss, success or failure, life or death. To keep track of all these variables, electronic computer/data processing equipment offers practical answers. This equipment is being used on an ever increasing basis to process large amounts of statistical, problematical, or experimental information, and to print out or display answers or information in very short periods of time. More and more reliance is being placed on the equipment to perform the repetitive, the experimental, and in some cases, even the whole programming operation for business, industry, government, and research groups.

Much has been written on the procedural steps required for study before installing electronic computer/data processing equipment. These requirements embrace selection of proper equipment, checking and planning for areas to receive the equipment, utility requirements, orientation and training of personnel to operate the equipment, as well as consideration for expansion of the initial facility. One other factor shall be included in this vital study — namely, protection against fires of either accidental or deliberate origin.

In addition to the hazards of fires from accidental causes, many computers and data processing installations have become prime targets for sabotage and arson.

Often times, the strategic importance placed upon electronic computer/data processing equipment by the user is vitally tied to uninterrupted operation of the system. Consequently, by the partial or entire loss of this equipment, an entire operation of vital nature could be temporarily paralyzed.

Not to be overlooked are the "one-of-a-kind" electronic computer/data processing systems. These are the "custom- made" models that are designed to perform specific tasks. Replacement units for this type of equipment are not available and the probability of the existence of duplicate facilities, which could be used to perform vital operations in the event the "one-of-a-kind" system is partially or totally impaired by a fire, is remote.

Computer equipment and materials for data recording and storage may incur damage when exposed to sustained ambient temperatures. The degree of such damage would be variable depending upon exposure, equipment design and the composition of materials for data recording and storage. The following are guidelines reflecting concernable sustained ambient temperatures:

1. Damage to magnetic tapes, flexible discs, and similar materials may begin at sustained ambient temperatures above 175°F (79.4°C), with the degree of damage increasing with further elevations of the ambient temperature and exposure time.
2. Damage to magnetic tapes, flexible discs, and similar materials may begin at sustained ambient temperatures above 100°F (37.8°C) and 120°F (48.9°C) can generally be reconditioned successfully where the chance of successful reconditioning lessens rapidly with elevations of sustained ambient temperatures above 120°F (48.9°C).
3. Damage to disc may begin at sustained ambient temperatures above 150°F (65.6°C) with the degree of damage increasing rapidly and further elevations of sustained ambient temperature.
4. Damage to paper products (including punched cards) may begin at a sustained ambient temperature of 350°F (176.7°C). Paper products which have not become brittle will generally be salvageable.
5. Damage to microfilm may begin at a sustained ambient temperature of 225°F (107.2°C) in the presence of steam or at 300°F (154.4°C) in the absence of steam.

Planning for fire protection is vital due to an organization's dependence upon the electronic computer/data processing equipment. Once management commits itself to a program of dependence on any such equipment, simple economics dictates doing away with former methods and procedures. The personnel, equipment, and facilities are no longer available to pick up the load assumed by the data processing equipment if it is put out of operation by fire or other unforeseen occurrences. Often the major cost involved to management by disruption of the computer operation is from business interruption rather than from the actual monetary loss represented by the equipment itself, although the latter may run into millions of dollars.

There are three major areas where judgment will be required in the application of this standard. They are:

1. Is this equipment important?
   This is a judicious evaluation based on both what the equipment is and what it does. If it controls air traffic safety it can be vital to human life; if it controls corporate information it can be vital to business "life," but if its loss would be simply inconvenient, then, perhaps, it is not important even though expensive.
2. Does this equipment need special construction?
   When special construction is needed in an important computer installation, it is essential that this construction of itself provide a safe environment for the equipment.
3. What is the exposure to the equipment?
   Exposure to destruction can come from within a computer cabinet, from within the equipment room, from the immediate area around the data processing room, from the

floors above and below the computer, and from outside of the building in which the equipment is located. This exposure can be evaluated and then controlled as needed.

The application of this standard to the protection of an individual system will depend upon the answers to these three questions.

While this standard cannot cover all contingencies with hard and fast rules, it does give an indication of the major areas of consideration requirements. There is no substitute for informed consideration of common sense principles.

The electronic computer area shall be located to minimize exposure to fire, water, corrosive fumes, heat, and smoke from adjoining areas and activities.

The computer room shall not be located above, below or adjacent to areas or other structures where hazardous processes are located unless adequate protective features are provided.

Computer Area Perimeter Construction and Fire Cut-offs and/or Remodelling

Where exposure to the building housing the computer is unfavourable, exposure protection shall be provided. This protection shall consist of blank masonry walls, or other suitable exposure protection, depending upon local conditions.

The computer room shall be cut off from other occupancies within the building by fire-resistant rated walls, floor and ceiling. The fire resistance rating shall be commensurate with the exposure, but not less than one hour.

The fire-resistant walls or partitions enclosing the computer room and/or storage rooms shall extend from the structural floor to the structural floor above, or the roof.

Computer Area Interior Construction

All materials used in construction within cut-off areas, including walls, floors, partitions, finish, acmustical treatment, raised floors, suspended ceilings, and other construction involved in the computer room, shall have a flame-spread rating of 25 or less. Except for resilient tiles or high-pressure plastic laminates may be used as the wearing surface on any exposed floors. Carpeting is permitted if it is demonstrated that the proposed material will not contribute to the spread of fire, and does not restrict lifting of panels for access to the under floor space.

Exposed cellular plastics shall not be used in computer room construction.

A structural floor on which a computer system is located, or which supports a raised floor installation shall incorporate provisions for drainage of the floor surface to minimize damage to the system and associated wiring due to domestic water leakage, sprinkler operation, coolant leakage, or firefighting operations.

Structural supporting numbers for raised floors shall be of concrete, steel, aluminium, or other non-combustible material.

Decking for raised floors shall be one of the following:

1. Concrete, steel, aluminium, or other non-combustible material, or
2. Pressure impregnated, fire-retardant treated lumber having a flame spread rating of 25 or less (see NFPA 255, Method of Test of Surface Burning Characteristics of Building

Materials), or wood or similar core material which is encased on the bottom with sheet, cast or extruded metal, with all openings or cut edges covered with metal or plastic clips or grommets so that none of the core is exposed, and has an assembly flame spread rating of 25 or less.

Existing combustible raised floors shall be replaced with construction meeting the requirements set forth by the authorities having jurisdictions or:

1. An automatic detection system shall be installed in the space below the raised floor. This additional automatic detection system shall meet the local requirements and shall sound an audible and visual alarm; and
2. Air space below the raised floor shall be subdivided into areas not exceeding 10,000 sq. ft. (929 sq. m) by tight non-combustible bulkheads.

Access sections or panels shall be provided in raised floors so that all the space beneath is easily accessible.

Electric cable openings in floors shall be made smooth or shall be otherwise protected to preclude the possibility of damage to the cables.

Opening in raised floors for electric cables or other uses shall be protected to minimize the entrance of debris or other combustibles beneath the floor. This may be accomplished by non-combustible covers, grilles, screens, or by locating equipment directly over the opening.

Cable openings or other penetrations into the computer area shall be sealed with material meeting the necessary requirements.

When the air space below a raised floor or above a suspended ceiling is used as a supply or return for air conditioning, the construction shall be non-combustible, and all wiring shall be of a listed type.

All office furniture in the computer room shall be of metal construction or of other materials that do not contribute significantly to the combustible contents.

The following shall not be permitted within the computer room:

1. Any activity or occupancy not directly associated with the electronic computer system(s) involved.
2. Supplies of combustibles such as paper, corrugated boxes, cards, inks, or equivalent printing materials, in excess of that needed for efficient operation.
3. Service and repair shops and operations except for that servicing and repairing performed directly on machines which are impractical to remove from the computer room.
4. Bulk storage of records.
5. Any other combustible material, equipment or operation constitutes a hazard and which can be removed.

Paper stock, unused recording media, and other combustibles within the computer room shall be restricted to the absolute minimum necessary for efficient operation. Any such materials in the computer room shall be kept in totally enclosed metal file cases or cabinets, or if provided for in individual machine design, shall be limited to the quantity prescribed by the equipment manufacturer.

The operation of an electronic computer system frequently requires considerable quantities of stationery supplies and other combustible support materials. This material can present a serious fire exposure within the computer room capable of causing serious damage to vital equipment or records.

Reserve stocks of paper, unused recording media, and other combustibles shall be stored in one or more rooms outside of the computer room.

Each individual unit shall be constructed so that by limiting combustible materials, or by use of enclosures, fire is not likely to spread beyond the unit in which the source of ignition is located. Enclosures of floor standing equipment having external surfaces of combustible materials of such size that might contribute to the spread of an external fire shall have a flame spread rating of 50 or less (see NFPA 255, Method of Tests of Surface Burning Characteristics of Building Materials). Automatic protection shall be provided for all units not so constructed.

Listed equipment shall be considered as meeting the necessary requirements.

Equipment meeting the requirements of U.L. Standard shall be considered as meeting the requirements.

Interconnecting cables and wiring between units, power cords, plugs, and connectors shall be of a listed type. They shall be considered as part of the computer system and suitable for installation of the floor or under a raised floor as describe above.

Approved flexible cord and plug assemblies used for connecting computer equipment to the branch circuit to facilitate interchange shall not exceed 15 ft. (4.57 m) in length.

Air filters for use in the cooling systems of individual units shall be a listed type that will not burn freely or emit a large volume of smoke or other objectionable products of combustion when attacked by flames. They shall be so arranged that they can readily removed, inspected, cleaned or replaced when necessary.

If the design of the unit is such that oil or equivalent liquid is required for lubrication, cooling or hydraulic purposes, it shall have a closed cup flash point of 300°F (149°C) or higher, and the container shall be of a sealed construction, incorporating automatic pressure relief devices.

All sound deadening material used inside of computer equipment shall be of such material or so arranged that it does not increase the potential of fire damage to the unit or the potential of fire propagation from the unit.

U.L. listed portable carbon dioxide, Halon 1301, or Halon 1211 extinguishers shall be provided and maintained for electrical fires.

U.L. listed Class A extinguishers of plain water type only shall be provided and maintained for use on fires in ordinary combustible materials, such as paper and plastics.

A sign shall be located adjacent to each portable extinguisher and shall plainly indicate the type of fire for which it is intended.

In installations where conditions may require the provision of carbon dioxide hand hose lines, they shall be installed and maintained in accordance with NFPA 14, Standpipe and Hose Systems.

In installations where conditions may require the provision of carbon dioxide hand hose lines, they shall be installed and maintained in accordance with NFPA 12, Carbon Dioxide Extinguishing Systems.

Automatic detection equipment shall be installed to provide early warning of fire. The equipment used shall be a listed smoke detection type. Each installation shall be engineered for the specific area to be protected, giving due consideration to air currents and patterns within the space and shall be installed and maintained in accordance with NFPA and local codes and standards.

Halon 1301 Total Flooding Systems

Where there is a critical need to protect data in process, reduce equipment damage, and facilitate return to service, consideration shall be given to the use of Halon 1301 total flooding systems in sprinklered or unsprinklered computer rooms.

In installations where Halon 1301 Total Flooding Systems are used, they shall be installed and maintained in accordance with the requirements of NFPA and local codes and standards.

Halon 1301 Systems shall be automatically actuated by an approved method of detection. To ensure detection, particular attention shall be given to the choice of actuation means, the air flows usually involved in such air handling systems, and the snail heat release under fire conditions.

The Halon 1301 System may be actuated by the automatic fire detection system meeting NFPA and local codes.

Where operation of the air conditioning system would exhaust the agent supply, the Halon 1301 System shall be interlocked to shut down the air conditioning when the Halon system is actuated. It is not necessary to de-energize the computer equipment prior to discharge.

Alarms shall be provided to give positive warning of a discharge or pending discharge.

An automatic sprinkler system shall be provided to protect the computer room when:

1. The computer room construction contains any combustible materials other than permitted in 2–4.1, or
2. The enclosure of a unit in a computer system, or the unit structure, is built all or in part of a significant quantity of combustible materials, or
3. The operation of the computer room involves a significant quantity of combustible materials.

Automatic sprinkler systems protecting computer rooms or computer areas shall be installed in accordance with NFPA 13, Installation of Sprinkler Systems.

To minimize damage in electronic computer equipment located in sprinkler protected areas, it is important that power be off prior to the application of water on the fire.

Air conditioning equipment shall conform to the requirements of NFPA, Air Conditioning and Ventilating Systems, and to the local codes and standards.

A separate air conditioning system should be provided for the computer area.

If the air conditioning system for the computer room also serves other areas, dampers to protect against both smoke and fire shall be provided in the duct work at all penetrations through the computer room fire cut-off.

Air ducts serving other areas shall not pass through the electronic equipment area, or suitable fire dampers shall be provided in the ducts as outlined in NFPA, Air Conditioning and Ventilating Systems.

All duct insulation and linings shall be non-combustible, including vapour barriers and coatings.

Air filters for use in air conditioning systems shall be listed types that will not burn freely or emit a large volume of smoke or other objectionable products of combustion when attacked by flames. They shall be so arranged that they can be readily inspected and cleaned or replaced when necessary.

The requirements in this section apply to all power and service wiring supplying the electronic computer equipment. They do not apply to wiring and components within the actual equipment or to wiring connecting various units of equipment.

Service equipment supplying the main power requirements of the computer room area shall be of a type arranged for remote control.

All wiring shall conform to NFPA 70, National Electrical Code.

If a service transformer must be installed in the computer area, it shall be of the dry type or the type filled with a non-flammable dielectric medium. Such transformers shall be installed in accordance with the requirements of NFPA 70, National Electrical Code. Service transformers should not be permitted in the electronic computer room.

Protection against lightning surges shall be provided where needed in accordance with the requirements of NFPA and National Electrical Code.

The number of junction boxes in under floor areas shall be kept to a minimum. Where used, they shall be metal completely enclosed, easily accessible, properly grounded and in compliance with the National Electrical Code requirements as to construction. They shall be securely fastened. No splices or connections shall be made in the under floor area except within junction boxes or approved type receptacles or connectors.

Emergency Power Controls

A disconnecting means shall be provided to disconnect the power to all electronic equipment in the computer room. This disconnecting means shall be controlled from locations readily accessible to the operator at the principal exit doors. There shall also be a similar disconnecting means to disconnect the air conditioning system servicing this area. Provision should be made for emergency lighting.

If a separate cooling system is required for operation of a computer installation, it shall be provided with a suitable alarm to indicate inadvertent loss of fluid.

38 Hand Held Extinguishers

The basic types of fires are Classes A, B, C, and D as in the following subsections.

Class A fires are fires in ordinary combustible materials, such as wood, cloth, paper, rubber, and many plastics.

Class B fires are fires in flammable liquids, oils, greases, tars, oil base paints, lacquers, and flammable gases.

Class C fires are fires which involve energized electrical equipment where the electrical nonconductivity of the extinguishing media is of importance. When electrical equipment is de-energized, extinguishers for Class A or B fires may be used safely.

Class D fires are fires in combustible metals, such as magnesium, titanium, zirconium, sodium, lithium, and potassium.

Portable fire extinguishers are classified for use on certain classes of fires and rated for relative extinguishing effectiveness at temperature of plus 70°F (21.1°C) by testing laboratories. This is based upon the preceding classification of fires and the fire extinguishment potentials as determined by fire tests.

The classification and rating system described in this standard is that used by Underwriters Laboratories Inc., and Underwriters' Laboratories of Canada and is based on extinguishing preplanned fires of determined size and description as follows;

Class A Rating — Hood and excelsior.

Class B Rating — Two-in. (5.1 cm) depth n-heptane fires in square pans.

Class C Rating — No fire test. Agent must be a non-conductor of electricity.

Class D Rating — Special tests on specific combustible metal fires.

The classification of extinguishers shall consist of a Letter which indicates the class of fire on which an extinguisher has been found to be effective preceded by a rating Numerical (Class A and B only) which indicates the relative extinguishing effectiveness. Except for Extinguishers classified for use on Class C or D hazards shall not be required to have a numeral preceding the classification letter. Portable extinguishers shall be maintained in a fully charged and operable condition, and kept in their designated places at all times when they are not being used.

Extinguishers shall be conspicuously located where they will be readily accessible and immediately available in the event of fire. Preferably they shall be located along normal paths of travel, including exits from an area.

Cabinets housing extinguishers shall not be locked.

Except where extinguishers are subject to malicious use, locked cabinets may be used provided they include means of emergency access.

Extinguishers shall not be obstructed or obscured from view. Except in large rooms, and in certain locations where visual obstruction cannot be completely avoided, means shall be provided to indicate the location.

Extinguishers shall be installed on the hangers or in the brackets supplied, mounted in cabinets, or set on shelves unless the extinguishers are of the wheeled type.

Extinguishers installed under conditions where they are subject to dislodgement shall be installed in brackets specifically designed to cope with this problem.

Extinguishers installed under conditions where they are subject to physical damage shall be protected from impact.

The selection of extinguishers for a given situation shall be determined by the character of the fires anticipated, the construction and occupancy of the individual property, the vehicle or hazard to be protected, ambient-temperature conditions, and other factors. The number, size, and placement of extinguishers required shall be determined by using related charts.

The minimum number of fire extinguishers needed to protect a property shall be determined as outlined in chapter for Job Site. Frequently, additional extinguishers may be installed to provide more suitable protection. Extinguishers having ratings less than specified may be installed provided they are not used in fulfilling the minimum protective requirements of this chapter.

Fire extinguishers shall be provided for the protection of both the building structure, if combustible, and the occupancy hazards contained therein.

Required building protection shall be provided by fire extinguishers suitable for Class A fires.

Occupancy hazard protection shall be provided by fire extinguishers suitable for such Class A, B, C, or D fire potentials as may be present.

Combustible buildings having an occupancy hazard subject to Class B and/or Class C fires shall have a standard complement of Class A fire extinguishers for building protection, plus additional Class B and/or Class C extinguishers. Where fire extinguishers have more than one letter classification (such as 2-A:20-B:C), they may be considered to satisfy the requirements of each letter class.

Fire Extinguisher Size and Placement for Class A Hazards

Minimal sizes of fire extinguishers for the listed grades of hazards shall be provided on the basis of Charts except as modified by local codes. Extinguishers shall be located so that the maximum travel distances shall not exceed those specified in charts, except as modified by local codes.

Certain smaller extinguishers which are charged with multipurpose dry chemical or Halo 1211 are rated on Class B and Class C fires, but have insufficient effectiveness to earn the minimum 1-A rating even though they have value in extinguishing smaller Class A fires.

Class B Fire Extinguishers for Pressurized Flammable Liquids and Pressurized Gas Fires.

Fires of this nature are considered to be a special hazard. Class B fire extinguishers containing agents other than dry chemical are relatively ineffective on this type of hazard due to stream and agent characteristics. Selection of extinguishers for this type of hazard shall be made on the basis of recommendations by manufacturers of this specialized equipment. The system used to rate extinguishers on Class B fires (flammable liquids in depth) is not applicable to these types of hazards. It has been determined that special nozzle design and rates of agent application are required to cope with such hazards. Caution: It is undesirable to attempt to extinguish this type of fire unless there is reasonable assurance that the source of fuel can be promptly shut off.

Fire Extinguisher Size and Placement for Class C Hazards. Extinguishers with Class C rating shall be required where energized electrical equipment may be encountered which would require a nonconducting extinguishing medium. This will include fire either directly involving or surrounding electrical equipment. Since the fire itself is a Class A or Class B hazard, the extinguishers are sized and located on the basis of the anticipated Class A or B hazard.

Electrical equipment should be de-energized as soon as possible to prevent reignition.

Extinguishers or extinguishing agents with Class D ratings shall be provided for fires involving combustible metals.

Extinguishing equipment shall be located not more than 75 ft. (22.7 n) from the Class D hazard.

Size determination shall be on the basis of the specific combustible metal, its physical particle size, area to be covered and recommendations by the extinguisher manufacturer on data from control tests conducted.

39 Carbon Dioxide Extinguishing Systems

For Carbon Dioxide to work properly as an extinguishing agent, a reasonably well enclosed space is in order to minimize the loss of the extinguishing medium. The area of allowable unlosable openings depends upon the type of combustibles involved.

For flash or surface type fires such as will be present with flammable liquids, any unlosable openings shall be compensated for by additional carbon dioxide as it is been specified.

If the quantity of carbon dioxide required for compensation exceeds the basic quantities required for flooding without leakage, the system may be designed for local application in accordance with NFPA and local codes.

For deep-seated fires such as will be involved with solids, unlosable openings shall be restricted to those bordering or actually in the ceiling.

To prevent fire from spreading through openings to adjacent hazards or work areas which may be possible resignation sources, such openings shall be provided with automatic closures or screening nozzles. The gas required for such protection shall be in addition to the normal requirement for total flooding. Where such confinement of gas is impracticable, protection shall be extended to include these adjacent hazards or work areas.

In the case of process and storage tanks where safe venting of flammable vapours and gases cannot be realized, the use of external local application systems is required.

Surface fires are the most common hazard particularly adaptable to extinguishment by total flooding systems. They are subject to prompt extinguishment when carbon dioxide is quickly introduced into the enclosure in sufficient quantity to overcome leakage and provide an extinguishing concentration for the particular materials involved.

For deep-seated fires, the required extinguishing concentration shall be maintained for a sufficient period of time to allow the smouldering to be extinguished and the material to cool to a point at which resignation will not occur when the inert atmosphere is dissipated. In any event, it is necessary to inspect the hazard immediately thereafter to make certain that extinguishment is complete and to remove any material involved in the fire.

The quantity of carbon dioxide for surface type fires is based on average conditions assuming fairly prompt extinguishment. A reasonable allowance for normal leakage is included in the basic volume factors but corrections shall be made for the type of material involved and any other special conditions.

Flammable Materials. Proper consideration shall be given to the determination of the design concentration of carbon dioxide required for the type of flammable material involved in the hazard. The design concentration is determined by adding a suitable factor (20 percent) to the minimum effective concentration.

The quantity of carbon dioxide for deep-seated type fires is based on fairly tight enclosures. The concentration shall be maintained for a substantial period of time, but not less than 20 min. Any possible leakage shall be given special consideration since no allowance is included in the basic flooding factors.

Combustible Materials. For combustible materials capable of producing deep-seated fires, the required carbon dioxide concentrations cannot be determined with the same accuracy-possible with surface burning materials. The extinguishing concentration will vary with the mass of material present because of the thermal insulating effects. Flooding factors have, therefore, been determined on the basis of practical test conditions.

The flooding factors have been found to provide proper design concentrations for the rooms and enclosures listed.

Flooding factors for other deep-seated fires shall be justified to the satisfaction of the authority having jurisdiction before use. Proper consideration shall be given to the mass of material to be protected because the rate of cooling is reduced by the thermal insulating effects.

The volume of the space shall be determined in accordance with proper charts. The basic quantity of carbon dioxide required to protect an enclosure shall be obtained by treating the volume of the enclosure by the appropriate flooding factor given in tables.

Additional quantities of carbon dioxide shall be provided to compensate for any special condition that may adversely affect the extinguishing efficiency.

The distribution system for applying carbon dioxide to enclosed hazards shall be designed with due consideration for the materials involved and the nature of the enclosure since these items may require various discharge tines and rates of application.

The minimum design rate of application shall be based on the quantity of carbon dioxide and the maximum tine to achieve design concentration.

For surface fires the design concentration shall be achieved within 1 min.

For high pressure systems, if a part of the hazard is to be protected by total flooding, the discharge rate for the total flooding portion shall be computed as specified.

For deep-seated fires the design concentration shall be achieved within 7 min but the rate shall not be less than that required to develop a concentration of 30 percent in 2 min.

Extended Rate of Application. Where leakage is appreciable and the design concentration must be obtained quickly and maintained for an extended period of time, carbon dioxide provided for leakage compensation may be applied at a reduced rate.

This type of system is particularly applicable to enclosed rotating electrical apparatus, such as generators, motors, and converters, but it may also be used on ordinary total flooding applications where suitable.

The minimum design concentration shall be obtained within the time limits and specified.

The extended rate of discharge shall be sufficient to maintain the minimum concentration.

For enclosed rotating electrical equipment a minimum concentration of 30 percent shall be maintained for the deceleration period, but not less than 20 min.

40 Dry Chemical Systems

Types of hazards and equipment for which dry chemical extinguishing systems shall be considered satisfactory protection include the following:

1. Flammable or combustible liquids and combustible gases. Extinguishment of uncontrolled discharge of flammable liquids or combustible gases may result in a subsequent explosion hazard.
2. Combustible solids having burning characteristics similar to naphthalene and pitch, which melt when involved in fire.
3. Flammable liquids, combustible liquids, or combustible gases released from transfer or loading facilities.
4. Electrical hazards such as transformers or loading facilities.
5. Textile operations subject to flash surface fires. Where bicarbonate base dry chemical is used, water shall be provided to extinguish possible smouldering or deep-seated fire.
6. Ordinary combustibles such as wood, paper, or cloth using multipurpose dry chemical when it can reach all surfaces involved in combustion.
7. Kitchen hoods, ducts, and associated range top hazards such as deep fat fryers. For other specific details, see Standard for the Installation of Equipment for the Removal of Smoke and Grease-Laden Vapour from Commercial Cooking Equipment.
8. Some plastics depending upon the type of material and their configuration of hazard. For more specific information, consult thc manufacturer of the equipment.

Types of materials for which dry chemical extinguishing systems shall not be considered satisfactory protection include the following:

1. Chemicals containing their own oxygen supply such as cellulose nitrate.
2. Combustible metals such as sodium, potassium, magnesium, titanium, and zirconium. Dry powder systems listed by a testing laboratory for combustible metal fires may be used.
3. Deep-seated or burrowing fires in ordinary combustibles where the multipurpose dry chemical cannot reach the point of combustion.

In total flooding systems where there is a possibility that personnel may be exposed to a dry chemical discharge, suitable safeguards shall be provided to ensure prompt evacuation of such locations, and also to provide means for prompt rescue of any trapped personnel. Such safety items as personnel training, warning signs, discharge alarms, predischarge alarms, and respiratory protection shall be considered. In local application systems where the dry chemical is likely to discharge upon personnel, such as in loading racks, discharge alarms and special personnel training shall be provided.

The discharge of large amounts of dry chemical may create hazards to personnel such as reduced visibility and temporary breathing difficulty.

Methods of Actuation. Systems shall be classified as automatic or manual in accordance with the following methods of actuation:

1. Automatic Operation. Operation that does not require any human action.
2. Normal Manual Operation. Operation of a system requiring human action where the device used to cause the operation is located near the hazard so as to be easily accessible at all tine. Operation of one control shall be all that is required to bring about the full operation of the system.
3. Emergency Manual Operation. Operation of the system by human means where the device used to cause operation is fully mechanical in nature and is located on the device being controlled or on its mounting assembly. "Fully mechanical" may incorporate use of the system pressure to complete operation of the device.

The amount of dry chemical in the system shall be at least sufficient for the largest single hazard protected, or for the group of hazards which are to be protected simultaneously.

The dry chemical used in the system shall be supplied by the manufacturer of the equipment. The characteristics of the system are dependent upon the composition of the dry chemical and the type of expellant gas, as well as upon other factors, and therefore, it is imperative to use the dry chemical provided by the manufacturer of the system.

41 Fire Protection for Nuclear Power Plant

Fire protection is both an art and a science. Perfection and practice in the art is the objective, with fire prevention being the ultimate goal. Fire prevention in the absolute sense is only possible where there is no combustible material to fuel a fire. The presence of combustible material creates a fire potential, no matter how slight. Thus, fire prevention, must as a first priority, consider the presence of any combustible material as a variance.

The total elimination of combustible material is seldom possible; therefore, fire protection requires additional measures to limit the consequences of fire.

A defense-in-depth philosophy of fire prevention, control and extinguishment shall be adopted and implemented to minimize and mitigate the effects of fire, and reduce hazards to personnel and property damage to acceptable minimums.

A well-balanced fire protection program includes prevention, detection, extinguishment of fires, safety to life, and preservation of property. The protection of the environment and of the public against nuclear hazards takes priority over that of the plant itself, and is addressed by the appropriate regulatory agencies. However, the size of nuclear power plants makes the economic impact of a forced outrage such that protection measures must be extended to include provisions to assure their continued operation.

Special Considerations in Fire Protection Encountered at Nuclear Power Plants. Consideration of the need for nuclear safety results in several areas of fire protection emphasis uniquely to the nuclear power plants, these are:

1. Reactor Shutdown Systems.
2. Cooling System Integrity.
3. Filtering System Integrity.
4. Ventilating System Integrity.

Nuclear power plants use the concept of defense-in-depth to achieve the high degree of safety required in the nuclear safety systems of the plant. This concept shall be extended to fire protection for the remaining areas of the plant.

With respect to the fire protection program the defense-in-depth principle is aimed at achieving an adequate balance in:

1. Preventing fires from starting.

2. Detecting fires quickly and suppressing those fires that occur, thereby limiting damage.
3. Designing the plant to limit the consequences of fire.

Fire Protection Management Program. The fire protection program for a nuclear power plant shall cover design features, personnel, procedures, plans, and equipment.

Senior management participation in this program shall begin with early design concepts and plant layout and continue through plant operation. In order to effectively develop and conduct this program, a fire protection manager shall be appointed at the conceptual stage of the project. He shall be selected on the basis of education, experience and advancement as an industrial fire protection engineer. He shall establish liaison with all internal departments, with all authorities having jurisdiction, and with the public fire department.

Safety to the lives of on-site personnel, protection of property, and continuity of power production as affected by the possibility of fires shall be given appropriate consideration in the general arrangement of the nuclear power plant. For the purposes of fire hazard analysis the plant component facilities shall be divided into primary facilities, secondary facilities, and support facilities.

The nuclear power plant will generally consist of the following functional facilities which would be expected to vary in their physical relationship to each other from plant to plant.

1. Typical Primary Facilities

   Reactor Building. Houses the reactor within a containment structure and its integral or free standing shield.

   Reactor Auxiliary Building. Contains reactor auxiliary and emergency core cooling system (ECCS) equipment.

   Control Building. Houses the main and auxiliary control panels, emergency switchgear, and batteries.

   Turbine Building. Houses the turbine generator and turbine generator auxiliaries.

   Stena Generator Building (LMFBR). Houses the steam generator.

   Intake Pumping Station. Houses the condenser circulating water makeup pumps, the raw service water and fire protection pumps, and all necessary valves and strainers for these systems.

   Electrical Switchyard. Encompasses the electrical transmission system coming into the site and leaving the site.

   Emergency Power Generation Facility. That facility designed to provide on-site emergency power in the event of an off-site power failure.

   Emergency Service water Pumping Station. Houses the emergency service water pumps, and all necessary valves, strainers, and electrical switchgear for these systems.

   Condenser Circulating water Pumping Station. Houses the condenser circulating water pumps, pump isolation valves pump suction and discharge conduits, and associated electrical equipment.

Electrical Control and Communications Building. Houses the switchyard relays, terminal communications equipment and may act as a control center for the switchyard up until the time the main control room becomes operation.

Main Heat Rejection System. Include cooling towers and facilities associated with spray ponds and canals.

Station Transformers. Those transformers which provide auxiliary power to the plant or transmit power from the plant.

2. Typical Secondary Facilities

Fuel Building. Inclusion of primary equipment in this building would require upgrading of the affected portions of the building to primary facility status.

Radwaste Building. Houses the equipment for processing radioactive waste from the plant.

Makeup Water Treatment Plant. Facility for producing high purity water before use in the reactor or its support systems.

Condenser Circulating Hater Treatment Building. Houses the plant and condenser circulating water biocide treatment system.

Demineraliser Regeneration Building. Contains facilities for regeneration and cleaning of resins from the plant condensate polishing system.

3. Typical Support Facilities

Office Building. Houses the offices for the plant's administrative employees and contains plant records.

Service Building. Encompasses the guard house, main gate, and office space for the plant's security force.

Auxiliary Boiler Building. Contains boilers for providing heating steam and steam to operate plant auxiliary equipment during the time that the nuclear steam supply is not available.

For these three types of facilities, safety to the lives of on-site construction and operating personnel and loss of property shall be given consideration as specified in Chapter 17 and elsewhere in this standard.

For primary and secondary facilities the cost of lost revenue and replacement power in addition to property loss shall be factored into the fire hazard analysis for determining the cost/benefit ratio for selecting the appropriate fire protection systems.

A fire hazard analysis shall be developed that will define the fire hazards that can exist and describe the loss limiting criteria to be used in the design of the facility. The fire hazard analysis shall take into consideration the basic data of the plant such as the functional subdivision of the facility, the controlled zones which have limited access, and the general plant layout of equipment or systems to develop the fire zone and the ratings of fire barriers.

Combustible materials in both large and small concentrations will be present in nuclear power plants, as in most other industrial plants, and it shall be assumed that outbreaks of fire occur for a variety of reasons.

For the purpose of assessing the fire loading, an inventory of all flammable and combustible materials shall be made for each area, identifying the location, type, quantity and form of the materials. The materials shall be classified into:

1. Flammable and Combustible Materials. Typical examples of flammable and combustible materials found in a nuclear power plant are:
   - Conventional fuels for emergency power units auxiliary boilers, etc.
   - Lubricants and oil.
   - Insulating materials (thermal and electrical).
   - Building materials (incl. PVC and other plastics).
   - Filtering materials (e.g., oil bath filters, charcoal, etc.).
   - Cleansing materials.
   - Paints and solvents.
   - Packaging materials (e.g., bitumen, etc.).
   - Neutron shields (if organic materials).
   - Clothing.
2. Flammable Gases. Typical examples of flammable gases found in a nuclear plant are:
   - Hydrogen for generator cooling, for coolant conditioning of gas-cooled reactors, and from battery charging. Propane or other fuel gases.
   - O_2 and H_2 by radiolysis in the core and addition of H_2 for improved recombination.
   - Gas for cutting and welding.
3. Combustible Radioactive Substances. Typical examples of radioactive substances external to the reactor are:
   - Sealed radioactive material, such as irradiated and/or plutonium containing fuel elements, irradiated control rods, neutron sources, etc.
   - Unsealed radioactive material, such as ion exchanger fillings and filter cartridges which have become loaded with radioactive substances, radwaste materials, etc.
4. Reactive Metals. Typical examples of reactive materials found in an LMFBR nuclear power plant are:
   - Sodium.
   - Sodium-potassium eutectic (NaK).

   While assessing the hazardous substances, ways and means of transporting the supplies of consumable goods on the site shall be considered.

   Temporary but predictable and repetitive concentrations of combustible materials shall also be considered. These may include:

   i. Replacement of lubricating or hydraulic oils.
   ii. Repainting equipment or structures.
   iii. Replacement of combustible filter materials.

iv. Scaffolding necessary to maintain or replace equipment.

v. Spare equipment in shipping crates or boxes awaiting installation.

Combustibles, other than those which are an inherent part of operation, shall be restricted to protected compartments or spaces.

As part of the protective measures, consideration shall be given to reducing the fire loading of contents in primary or secondary facilities by reducing the amount of combustible materials, wherever possible, as indicated in the following design features:

1. Provision of separate piping systems for the lubricating system and control system of the turbogenerators.
2. Use of an approved low hazard synthetic hydraulic fluid in the control systems.
3. Use of air circuit breakers, low oil content circuit breakers, or sulphur-hexafluoride circuit breakers.
4. Use of approved non-combustible insulation materials on components.
5. Use of air filters meeting the requirements of UL Class 1 and HEPA filters UL 586 (charcoal filters excepted).

Combustible liquid storage greater than 1100 gallons (4164 litres) shall be located outside of the building unless enclosed by a three-hour fire barrier.

The storage and use of hydrogen shall be in accordance with NFPA 50A.

Flammable and combustible liquid and gas piping shall be in accordance with ANSI and appropriate NFPA codes.

Auxiliary gas or oil-fired boilers shall comply with NFPA 85, Single Burner Boiler-Furnaces.

Construction materials for nuclear power plants shall be classified by at least, one of the following tests:

1. NFPA 255, Surface Burning Characteristics of Building Materials (UL 723 & ASTM E-84).
2. NFPA 251, Fire Tests of Building Construction and Materials (ASTM E-119).

Certain materials of construction that do not meet the definition of non-combustible or limited combustible may have to be used in a nuclear power plant, and thus the location in the plant, the amount exposed to ignition, and the proposed fire protection system shall be justified in writing and the approval of the plant fire protection manager and the authority having jurisdiction shall be obtained.

Wall and structural components, thermal insulation materials, radiation shielding materials, and soundproofing shall be non-combustible, limited combustible or listed by a nationally recognized testing laboratory for flame spread, smoke, and fuel contribution, of 25 or less in its use configuration.

Suspended ceilings, including light diffusers, and their supports shall be of non-combustible or limited combustible construction. Electrical wiring above suspended ceilings shall be in metallic conduit or solid bottom, solid covered ferrous raceways. Otherwise concealed spaces shall be devoid of combustibles.

Roof coverings shall be Class A as determined by tests describe in NFPA 256, Standard Methods of Fire Tests of Roof Coverings. Metal roof deck construction shall be Class I (Factory Mutual Approved) or "fire-acceptable" (as tested by Underwriters Laboratories Inc.).

Transformers installed inside of buildings shall be either dry type or insulated and cooled with an approved liquid, unless installed in vaults in accordance with Article 450 (C) of NFPA 70-1978, National Electrical Code.

Combustible cooling towers shall be protected with deluge sprinkler systems in accordance with NFPA 214, Water Cooling Towers. Fire hydrants shall be operational at the cooling tower site prior to beginning construction activities.

Since cooling towers can be vital to the operation of the plant and reconstruction can be lengthy, non-combustible construction should be used.

The nuclear power plant shall be subdivided into separate fire areas to minimize the risk of spread of fire and the resultant consequential damage from corrosive gases, fire suppression agents, smoke, and contamination from radioactive substances.

In multi-unit plants, each unit shall be separated from the other either by distance or fire barriers. The distance or fire barrier rating shall be determined from the fire hazard analysis.

Specific Fire Separation Requirements

An approved fire rated barrier of three-hours fire resistance, with equivalent rated automatic fire doors, and with approved penetration protection, shall be provided as follows:

1. To separate all contiguous buildings, such as turbine, reactor containment, auxiliary fuel handling, control, radwaste, service, administration, and other areas as dictated by reactor design and fire hazard analysis.
2. To isolate turbine generator lube oil conditioning or system room and lube oil storage tank from turbine buildings and adjacent areas.
3. To isolate emergency diesel generator rooms or buildings (emergency power generating areas) from each other and adjacent plant areas.
4. To separate diesel fire pumps from other pumps in the same pump house.
5. To separate all areas with concentrations of cables, such as cable spreading rooms, cable tunnels, cable penetration areas, cable shaft's or chases, included within the reactor containment, from adjacent areas.
6. To isolate auxiliary boiler rooms from adjacent areas.

Where vertical fire barriers are required, stairwells, elevators, and trash chutes shall be enclosed in an approved fire rated enclosure of two-hour construction with approved automatic or self-closing fire doors.

Buildings shall be protected from exposure fires involving oil filled transformers, by locating the transformer facilities at least 50 feet (15.2 m) from other facilities or by providing a two-hour fire barrier between transformers and exposed facilities. A one-hour fire barrier or a distance of 30 feet (9.1 m) shall be provided between adjacent transformers. Means shall be provided to contain oil spills.

Ventilation of a nuclear power plant involves balanced air differentials between plant areas, comfort ventilation, and heat removal from areas where heat is generated by equipment. This need also includes fire area isolation and smoke removal equipment, as well as equipment for filtering radioactive gases. The design of ventilation systems is further complicated by the seismic, tornado, and missile criteria for building penetrations.

NFPA Standards are the basic standards that shall be followed for the design installation and operation of the ventilation systems necessary for normal and emergency operation of the plant except as modified below.

General Requirements

Smoke and Corrosive Gas Removal

Automatic shutdown of ventilation systems by temperature or smoke detectors as prescribed by applicable NFPA standards in the following building shall be consistent with nuclear safety and safety of on-site personnel:

1. Reactor building.
2. Auxiliary building.
3. Control building.
4. Turbine building.
5. Intake pumping station.
6. Emergency service water pumping station.
7. Condenser circulating water pumping station.
8. Electrical control and communications building.
9. Fuel building.
10. Emergency power generating building.

Smoke, corrosive gases and the nonradioactive substances which might be freed by the fire shall be vented from their place of origin directly to a safe location. Radioactive materials that might be released by the fire shall be confined, removed from the exhaust ventilation air stream or released under controlled conditions.

Ventilation systems designed to exhaust smoke or corrosive gases shall be evaluated to ensure the inadvertent operation or failures will not violate the controlled areas of the plant design.

Smoke ventilation shall be provided for fire areas based upon the fire hazard analysis.

Separate smoke ventilation systems are preferred; however, smoke venting can be integrated into normal ventilation systems using automatic or manually positioned dampers and motor speed control.

The lack of smoke and heat venting in areas of relatively high combustible loading can result in significant damage to structural components.

Automatic or manual actuation of smoke and heat venting will be determined by the fire hazard analysis.

The fresh-air supply intakes to all areas shall be located remotely from the exhaust air outlets and smoke vents of other fire areas to minimize the possibility of contaminating the intake air with the products of combustion.

Enclosed stairwells shall be designed to minimize smoke infiltration during a fire.

Stairwells serve as escape routes and firefighting access routes. A suitable method of ensuring a smoke-free stairwell is using a slight positive pressure for the stairwell ventilation system.

When natural-convection ventilation is used, a minimum ratio of vent area to floor area shall be on to 200 except in oil hazard areas where a one to 100 ratio shall be provided.

When mechanical ventilation is used, 300 cfm (8.5 m^2/min) is equal to one square foot (0.09 m^2) of natural convection vent area.

To prevent corrosion which might be caused by direct release of chemicals in case of fire, the acids and alkalines used for the primary coolant treatment plant and stored in appreciable quantities on the site shall be protected so as not to increase the risk of damage in case of a fire.

The power supply and controls for mechanical ventilation systems shall be located outside the fire area served by the system or protected from fire damage.

Plastic ducts, including fire retardant types, shall not be used for ventilating systems.

Ventilation ducts that pass through fire areas which they do not serve shall not degrade the fire integrity of the fire rated enclosure.

Fire dampers or fire doors compatible with the rating of the barrier may be required at the duct penetrations to the fire area.

Fire dampers equal in rating to the fire barrier shall be provided to prevent the passage of smoke, heat, or flame through ventilation ducts from one fire area to another, except in locations where shutdown of the ventilation system is not allowed.

Two 1½ hour rated dampers are considered acceptable where a three-hour rating is required. Fire dampers shall be equipped with thermal elements. The closure of fire dampers shall be guaranteed by mounting the damper directly into the separating wall or by protecting the duct up to the damper according to the fire resistance of the separating wall structure.

Interconnections of individual fire areas via the ventilation system shall be avoided insofar as possible. Where this is not possible, the necessary precautions shall be taken to prevent the spread of smoke and fire by such routes.

Fire dampers in the interconnecting ventilation ducts should be provided when the ventilation system cannot be sectioned off in the normal manner.

False floors or suspended ceilings shall not be employed as common pressure equalizing chambers for redundant ventilation systems but may be used for distribution of air to the corresponding room.

Air entry filters shall have non-combustible filter media. They shall produce a minimum amount of smoke (UL Class 1) when subjected to heat. In order to decrease the fire hazard of these filters and of oil bath type filters, only approved fire-resistive adhesives and oils with an open-cup flash point equal to or greater than 464°F (240°C) and which do not produce appreciable smoke shall be used. HEPA filters shall meet UL Standard.

Fire suppression systems shall be installed to protect filters which collect combustible material, unless the elimination of such protection is justified by the fire hazard analysis.

Administrative Procedures and Controls

This section provides criteria for development of administrative procedures and controls necessary for the execution of the fire prevention and protection activities and practices for the operating plant. Included herein are the minimally accepted actions required of cognizant plant management to assure the performance of fire protection systems and personnel and the compliance with the fire prevention program.

The plant manager or his delegated fire protection manager shall be responsible for the periodic update of the fire hazard analysis.

The plant manager or his delegated fire protection manager shall be responsible for the following:

1. Develop, implement, and periodically update as necessary a fire brigade plan in accordance with Chapter 14.
2. Maintain housekeeping in such a manner so as to minimize the probability of fire causing loss of life or property damage.
3. Develop, implement, and periodically update as necessary an emergency evacuation plan for all personnel.
4. Develop, implement, and periodically update as necessary a welding and cutting safety procedure using NFPA 51B and NFPA 241 as guides (see Chapter 15).
5. Restrict smoking and other sources of ignition to properly designated and supervised safe areas of the plant.
6. Develop, implement, and periodically update as necessary a fire prevention surveillance plan integrated with periodic recorded rounds to all accessible unattended section of the plant.
7. Coordinate the periodic testing of all systems and equipment affecting fire prevention and fire protection in accordance with the applicable NFPA standards and/or the manufacturers' installers' instructions and procedures to include maintenance of appropriate documentation.
8. Develop an alternate protection plan for those instances when it becomes necessary to remove any fire protection equipment or system from service.
9. Fire investigation and the coordination of all plant fire reporting. Responsibilities including reviewing fire reports, taking corrective action, and making the proper distribution of the reports.
10. Conduct periodic inspection of the plant. A prepared checklist shall be used for the inspection. The areas of primary containment and high radiation areas normally inaccessible during plant operation shall be inspected as plan conditions permit but at least during each refuelling outage. The results of each inspection shall be documented and retained for a period of two years.

Plant administrative procedures shall specify appropriate requirements governing the storage, use, and handling of flammable liquids and gases.

Plant administrative procedures shall specify appropriate requirements governing the control of electrical appliances, i.e., portable electrical heaters in critical areas.

A signalling system shall be provided an designed with the consultation of the fire protection manager and shall be based upon the fire hazard analysis. The signalling system shall be selected from and comply with the NFPA standards.

The signalling system shall provide emergency operation for fire detection, fire alarm, and water flow alarms during a single break or single ground fault of the signalling line circuit.

Fire signalling systems shall give audible and visual alarm and annunciation in the control room. Local alarms shall be provided in the affected fire zone.

Audible fire detection alarms shall be distinctive and unique. They shall not be capable of being confused with any other plant system alarms.

Fire detection, alarm, and actuation systems shall be connected to an emergency power supply.

Automatic fire detectors shall be selected, installed, and maintained in accordance with NFPA standards.

Manual fire alarm boxes (pull boxes) shall be installed on the basis of need by the fire hazard analysis.

The fire control panel shall contain, as a minimum, the following readouts:

1. Visual alarm.
2. Audible alarm.
3. Indicators for control functions directly initiated by fire detection or protection devices.
4. Control instrumentation for fire suppression systems.
5. Uninterruptible or emergency power supply condition.
6. Data acquisition system for recording of signals.

Communications

The plant public address system shall be available on a priority basis for fire announcements, directing plant fire brigades, and evacuation information.

Radio communications shall be provided to back up the public address system.

Alarm system testing shall be performed and a written report shall be submitted to the plant manager and to the plant fire protection manager. The NFPA testing requirements shall be the minimum requirements.

Automatic fire suppression systems shall be provided in all areas of the plant as required by the fire hazard analysis. Except as modified in this chapter, the following NFPA standards shall be used.

1. NFPA 11, Foam Extinguishing Systems
2. NFPA 11A, High Expansion Foam Systems
3. NFPA 11B, Synthetic Foam and Combined Agent Systems
4. NFPA 12, Carbon Dioxide Systems

5. NFPA 12A, Halo 1301 Systems
6. NFPA 12B, Halon 1211 Systems
7. NFPA 13, Installation of Sprinkler Systems
8. NFPA 15, Water Spray Fixed Systems
9. NFPA 16, Foam-Water Sprinkler and Spray Systems
10. NFPA 17, Dry Chemical Systems.

The extinguishing systems chosen shall be based upon the design parameters required as a result of the fire hazard analysis. When the fire hazard analysis shows that a fixed extinguishing system is needed, Table 10-1.1 shall be used in selecting the type of system or combination of systems to be installed.

Selection of extinguishing agent shall be based upon:

1. Type or class of hazard.
2. Effect of agent discharge on critical equipment such as thermal shock, continued operability, water damage, over-pressurization, cleanup, etc.
3. Health hazards.

A designated person working in the plant shall be trained in the maintenance, inspection, operation and emergency actuation of all fixed fire suppression systems installed in the plant whether automatic or manual.

Each fire suppression system shall be equipped with approved alarm devices, annunciated locally or in the hazard area, and in the control room. Alarm circuits to the control room shall be installed and supervised in accordance with NFPA codes.

All shut-off valves controlling fire suppression systems shall be supervised by one of the following methods:

1. Electrical supervision with an audible signal in the control room.
2. Locking valves open.

Automatic Sprinklers and Water Spray Systems

Where water is used as an extinguishing agent, systems shall be designed, installed, and maintained in accordance with the NFPA codes. Hydraulically calculated are preferred.

Each system shall have an independent connection to the plant yard main and be equipped with an approved indicating type control or shut-off valve.

Multiple sprinkler and standpipe systems may be supplied by interior headers or fire protection loops. When provided, such headers or loops are considered an extension of the yard main system and should be provided with at least two connections to the yard main. The arrangement should be supplied and valved so that no single impairment can affect sprinkler and hose protection at the sane time.

Automatic sprinkler and water spray systems shall not be designed to protect more than one fire area.

Drainage of all buildings shall be assessed to assure safe run-off or retention of fire protection water and to minimize damage.

Equipment vulnerable to water damage should be placed on pedestals and floor openings should be curbed or sealed.

The installation of a completely, properly engineered automatic sprinkler or water spray system in a building will generally reduce the draining and retention tank requirements, as compared to a building which relies primarily on manual firefighting.

Liquefied Compressed Gas Suppression Systems

In areas where appreciable Class C hazards exist and where the use of other agents might result in equipment malfunction or damage as determined by the fire hazard analysis, liquefied compressed gas extinguishing agent systems shall be installed. These systems, where required, shall be designed, installed, and maintained in accordance with the appropriate agent standard as follows:

1. Halon 1301 — NFPA 12A
2. Halon 1211 — NFPA 12B
3. Carbon dioxide — NFPA 12.

When liquefied compressed gas extinguishing agent systems are used, they shall be automatically actuated by an approved method of detection meeting the requirements of NFPA 72E. To ensure rapid detection, particular attention shall be given to the choice of actuation means, the air flows usually involved, and the heat release rates for the hazard under fire conditions.

For carbon dioxide total flooding systems that are automatically actuated in normally occupied areas, a predischarge alarm and tine delay are required to allow for personnel evacuation of the area.

Halon 1211 is not acceptable for total flooding use in normally occupied areas.

Electrical equipment need not be de-energized prior to the discharge of these extinguishing agent systems, but shut-down is desirable if it can be accomplished.

Foam Systems

Where fires involving flammable liquids and gases under pressure are likely, foam systems shall not be used. These systems, where required by the fire hazard analysis, shall be designed, installed, and maintained in accordance with the appropriate agent standard as follows:

1. Foam Extinguishing Systems — NFPA 11
2. High Expansion Foam Systems — NFPA 11A
3. Synthetic Foam and Combined Agent Systems — NFPA 11B
4. Foam Water Sprinkler and Spray Systems — NFPA 16.

Foam systems may be used in areas where appreciable Class B hazards exist, especially where flammable liquid fuel in- depth or spills are likely.

High expansion foam systems are particularly suited for total flooding applications, but should not be used in normally occupied areas.

Foam systems shall be designed to be automatically actuated by an approved method of detection meeting the requirements of NFPA 72E.

Foam systems shall not be used to protect hazards involving energized electrical equipment.

Dry Chemical Systems

Where appreciable Class B hazards exist, dry chemical systems shall be considered as appropriate protection.

Dry chemical systems, utilizing multipurpose dry chemical, are also effective on Class A hazards, provided that "deep- seated" fires are not likely.

Dry chemical systems may be used in conjunction with AFFF in a combined agent application, where securing action is required in addition to extinguishment.

Electrical equipment need not be de-energized prior to system discharge, but shut-down is desirable if it can be accomplished.

Dry chemical systems are not recommended for use in hazards areas where sensitive electrical equipment is located, due to the potential for damage to the equipment from the dry chemical and its products of decomposition where moisture and high temperatures are likely.

Where dry chemical systems are required by the fire hazard analysis, they shall be designed, installed, and maintained in accordance with the requirements of NFPA 17.

Dry chemical systems shall be designed to be automatically actuated by an approved method of detection meeting the requirements of NFPA 72E.

The water supply system for the permanent fire protection installation shall be determined by the fire hazard analysis.

A water supply shall be provided during the initial construction period as specified in Section 16-3(b).

The fire main system piping shall not serve service water system functions. The fire water loop shall meet the construction requirements of NFPA 24, Outside Protection.

The water supply for the permanent fire protection installation shall be based on the maximum automatic sprinkler or fixed water spray system demand, with simultaneous flow of 750 gpm at grade (2835 litres/min) for hose streams and the shortest portion of the fire loop main out of service.

The use of multiple approved fire pumps shall be based on supplying the demand required with the largest pump out of service, utilizing different power sources.

Fire pumps shall meet the requirements of NFPA 20, Centrifugal Fire Pumps, and shall be automatic starting.

An automatic pressure maintenance pump (jockey pump) or a head tank shall be provided to keep the fire water main pressure at approximately 10 psig (68.9 kPa) above the start pressure setting of the fire pump.

Pumps shall take suction from acceptable sources of water. If the source is from tanks, at least two tanks shall be provided and each tank shall be sized to contain a minimum 2-hour fire water flow demand. Where tanks are used, these tanks shall be automatically filled from a source capable of providing a 2-hour supply for the fire protection requirement in 8 hours. Fire water supply tanks shall comply with NFPA 22, including freeze protection requirements.

The 8-hour requirement for refilling may be reduced or eliminated if the initial supply exceeds the 2-hour supply requirement on a volume/ratio basis.

Salt and tidal water shall not be acceptable primary natural water sources.

The following supervisory signals, where applicable, shall be received in the control room

1. Pump sunning
2. Power failure
3. Failure to start
4. Hater level
5. Pump room and tank temperatures.

Hand Portable and Wheeled Fire Extinguishers

Hand portable and wheeled fire extinguishers shall be installed, inspected, maintained, and tested in accordance with NFPA 10, and approved or listed by a recognized testing laboratory.

Hand portable and wheeled fire extinguishers shall be selected for the specific class or classes of hazards using the following descriptive guide.

Hand portable and wheeled fire extinguishers for protecting Class "A" hazards shall be selected from the following: water types, foam, Halon 1211, or multipurpose dry chemical.

Hand portable and wheeled fire extinguishers that contain water, water/antifreeze, or foam, should not be used on energized electrical equipment because a personal electrical shock hazard would be present.

Multipurpose dry chemical is not recommended for installation in locations having sensitive electronic equipment, due to the potential for damage to this equipment from the dry chemical and its decomposition products.

Hand portable and wheeled fire extinguishers for protection of Class "B" hazards shall be selected from the following: dry chemical types, Halon 1211, foam or carbon dioxide.

In addition to it extinguishing capabilities, foam can be utilized to secure a flammable liquid spill before ignition.

Some of the extinguishing agents, i.e., carbon dioxide and foam, are relatively ineffective on fires involving flammable liquids under pressure.

Hand portable and wheeled fire extinguishers for protecting Class "C" hazards shall be selected from the following: Halon 1211, carbon dioxide, dry chemical types.

Carbon dioxide units equipped with metal horns should not be used on fires in energized electrical equipment.

Dry chemical is not recommended for use in an area where sensitive electrical equipment is located, due to the potential for damage to this equipment from a dry chemical and its products of decomposition.

Hand portable and wheeled fire extinguishers and extinguishing agents for the protection of Class "D" hazards shall be of types approved for use on the specific combustible metal hazard.

Where approvals are not available for a specific combustible metal, recommendations of a manufacturer may be utilized providing proper test data or use records are available.

Agent for Class "D" hazards can be applied utilizing a scoop and a container.

A sign will be located adjacent to each hand portable and wheeled fire extinguisher and shall plainly indicate the type of fire for which it is intended.

All persons working in an area shall be thoroughly trained in the use of all types of hand portable and wheeled fire extinguishers in the area and shall be familiar with their location. This training shall include both the capabilities and limitations of each available type of extinguisher.

Hand Hose Lines

Hand hose lines utilizing water, foam, carbon dioxide, or dry chemical shall be installed, inspected, and maintained in accordance with the following NFPA standards, respectively: NFPA 14, NFPA 11, NFPA 11B, NFPA 12, and NFPA 17.

Hand hose lines utilizing water, foam, carbon dioxide, and dry chemical shall be selected for the specific class or classes of hazards utilizing the following descriptive guide.

Class "A" Hazards. Hand hose lines for the protection of Class "A" hazards shall be selected from the following: water, foam, and dry chemical (multipurpose only). The nozzles for these hand hose lines shall be approved, electrically safe nozzles in the plant areas where Class "A" hazards involve energized electrical equipment.

Class "B" Hazards. Hand hose lines for the protection of Class "B" hazards shall be selected from among the following: water, foam, and dry chemical.

Foam is not considered suitable alone for fires involving gases and liquefied gases nor for three-dimensional flammable liquid fires. If these hazards are encountered, hand portable, wheeled dry chemical fire extinguishers or dry chemical hand hose lines are most effective especially when used in combination.

Class "C" Hazards. Hand hose lines for the protection of Class "C" hazards shall be selected from: water, carbon dioxide, and dry chemical. Approved nozzles shall be provided for use on Class "C" hazard fires.

The use of hand hose lines involving water on energized electrical equipment shall take into consideration:

1. The voltage of the equipment.
2. The type of water stream used, i.e., solid vs. spray.
3. The effect of electrically conductive water on the operability of the equipment.
4. The most probable distance between the fire fighter and the energized electrical equipment.
5. The location of energized electrical equipment in the drainage path that is likely to be encountered when water is used.

Where sensitive electrical control equipment is involved, due consideration shall be given to the effects of thermal shock on the operability of this equipment when using carbon dioxide hand hose lines.

Carbon dioxide hand hose lines shall not be used in normally inhabited enclosures unless provisions for evacuation (alarm, annunciation, etc.) are made before use.

Class "D" Hazards. Hand hose lines involving water, foam, and carbon dioxide shall not be used for the protection of Class "D" hazards.

A sign shall be located adjacent to each hand hose station and shall plainly indicate the type of fire for which it is intended.

Designated persons working in the area shall be thoroughly trained in the inspection, location, and use of all hand hose line devices. This training shall include the capabilities and limitations for each type of hose station.

Detailed action plans shall be worked out by the fire protection manager for proper deployment of in-plan firefighting personnel. The plan shall require testing to verify its practicability and completeness by means of semi-annual fire drills. Such drills shall be documented and kept on file at the plant for a minimum period of one year. The action plan shall consider the need for coordination between private and public firefighting forces during drills and emergencies.

Involvement of state and local agencies in semi-annual drills is encouraged.

Standard for Fire Defense Measures. All fire protection systems and devices such as automatic fixed fire extinguisher systems, automatic fire alarm and various layout arrangements, i.e., fire doors, fire dampers, ventilation arrangements, smoke venting, etc., shall be subjected to recurring and documented checks and tests. The staff shall be trained and drilled so that each person is capable of acting quickly and efficiently in the event of fire or danger of fire. A senior management representative, along with the fire protection manager, shall be responsible for ensuring that the plan fire defense is the best possible in all respects, including that all firefighting equipment and systems are in working order. This supervisory activity shall be done by developing a plan emergency plan.

Liaison with Public Fire Authorities. Liaison between and interfacing with the fire protection manager and the public fire authorities shall be maintained. It shall be the duty of the fire protection manager, together with the plant manager and plant security manager, in consultation with the public fire department, to make plans for firefighting and rescue, including assistance from other organizations, and to maintain these plans in a current mode. Fire fighters from the public fire department and from other fire departments, including other industrial fire brigades which may be expected to respond to a fire at the plant, shall be familiarized with the plan layout. The access routes to fires in the controlled area (to which access doors are locked) shall be planned in detail in advance. The public fire department shall be given instructions and trained how to react in areas where radioactive materials, radiation, or hazardous materials may be present.

Regular Training and Drills. Fire brigade members shall be given regular training and practice in firefighting and rescue routines, including radioactivity and health physics considerations, to ensure that each member is thoroughly familiar with the steps to be taken in the event of fire, which will contribute to maintaining the best possible preparedness for such contingencies. Each fire brigade member shall be capable of acting:

1. To alarm the plant fire brigade and public fire department.
2. To identify any alarm zone or fire protection system that has operated.
3. To use available rescue and extinguishing devices.

4. To actuate the fixed fire extinguishing systems, the smoke venting systems, etc.
5. To cooperate with and assist the public fire department in its work.
6. To train plant personnel in room entry procedures where total flooding fire suppression systems are used.

Equipment for the Plant Fire Brigade

The conventional equipment for the plant fire brigade shall be selected to suit the entire plant site. To complement the conventional equipment the following equipment shall be provided to cope with the nuclear hazard:

1. Protective clothing.
2. Respiratory protective equipment.
3. Radiation monitors.
4. Personal dosimeters.
5. A formal program for the care and maintenance of the fire brigade equipment shall be established by the owner through the fire protection manager.

42 Cooling Tower

Cooling tower is a device for reducing the temperature of a liquid, usually water, by bringing it into contact with an air stream where a small portion of the liquid is evaporated and the major portion is cooled. Cooling towers are used for large-scale air-conditioning installations, in the chemical industry, for cooling steam condensers of inland power plants if insufficient lake or river water is available, and in a number of other industrial applications.

Atmospheric air is normally not saturated with water and can absorb additional moisture until its relative humidity reaches 100 percent. This absorption process can be aided if the water is broken up into many small droplets, increasing the water surface area and thus the evaporation. The energy required for the evaporation into the airstream is provided by a fall in the temperature of the remaining water. For each pound of water evaporated, about 980 BTU of energy become available for cooling the remaining water and for heating the airstream, an undesirable but necessary side effect [See British Thermal Unit (BTU)]. In practice, cooling towers work with water losses from 2 to 5 percent, depending on the conditions of operation.

The hot water is initially piped to the top of the cooling unit and the broken up into droplets by being passed through a perforated plate. In large towers, a series of such plates are arranged on top of each other; the water drips down and comes in contact with upward flowing air. In small towers, the water simply drops over a series of slats. The tendency for the heated air to rise provides the airflow for very small cooling towers; but most larger units depend on having air blown over the plate arrangement, either by induced draft. with a suction fan on the top or by forced draft. with a blower on the bottom.

Cooling towers have become part of the nuclear power controversy of the 1980s. Although they add greatly to the building costs of nuclear installations, the alternative to cooling towers, the discharging of large volumes of heated water into lakes and rivers, has been proved detrimental to plants and animals in lakes and rivers and along their shores.

The following are some of the factors which shall be considered in determining the extent and method of fire protection of induced draft and natural draft cooling towers:

1. Importance to continuity of operation.
2. Size and construction of tower.
3. Type of tower.

4. Location of tower.
5. Hater supply.
6. Value of tower.

Depending on factors indicated above where a fire protection system is required, one of the following general types of systems shall be used:

1. Open head deluge system.
2. Closed head dry-pipe system.
3. Wet-pipe automatic sprinkler system.

Types of Systems

The counterflow tower design lends itself to either closed or open head systems. Therefore, wet-pipe, dry-pipe, or deluge systems may be used. Where water supplies are adequate, the deluge system provides a higher degree of protection.

The crossflow design is such that it is difficult to locate sprinklers in the most desirable spots for both water distribution and heat detection. This situation can be solved by separating these two functions and using separate water discharge and detection systems. The open head deluge system does this and, therefore, is the type to be used in crossflow towers.

Minimum Rate of Application

- Under the fan decks of counterflow towers the rate of application of water shall be 0.5 gpm/sq ft. (20.4 L/min/m^2) including fan opening).
- Under the fan decks of crossflow towers, the rate of application of water shall be 0.33 gpm/sq. ft. (13.45 L/min/m^2) (including fan opening).
- Over the fill areas of crossflow towers, the rate of application of water shall be 0.5 gpm/ sq. ft. (20.4 L/min/m^2).

Types and Locations of Discharge Outlets.

In counterflow towers, the discharge outlets shall be located under the fan deck and fan opening.

In crossflow towers, the discharge outlets protecting the plenum area shall be located under the fan deck and in the fan opening. Discharge outlets protecting the fill shall be located under the distribution basin on either the louver or drift eliminator side, discharging horizontally through the joist channels. On towers with a fill air travel greater than 18 ft. (5.5 m), the outlets shall be located on opposite sides of the basins, in alternate joist channels. Where joist channels are wider than 2 ft. (.6 m), more than one discharge device may be required per joist channel.

On towers having extended fan decks which completely enclose the distribution basin, the discharge outlets protecting the fill area shall be located over the basin, under the extension of the fan deck. These discharge outlets shall be open directional spray nozzles arranged to discharge 0.35 gpm/sq. ft. (14.26 L/min/m^2) directly on the distribution basin, and 0.15 gpm/sq. ft. (6.11 L/min/m^2) on the underside of the fan deck extension.

Each system shall be provided with a shutoff valve so located as to be readily accessible during a fire in the area the system protects or adjacent areas, or, for systems installed for fire prevention, during the existence of the contingency for which the system is installed.

Automatically controlled valves shall be as close to the hazard protected as accessibility during the emergency will permit, so that a minimum of piping is required between the automatic valve and the spray nozzles.

Remote manual tripping devices, where required, shall be conspicuously located where readily accessible during the emergency and adequately identified as to the system controlled.

Strainers are required for systems utilizing nozzles with waterways less than nominal 3/4 in. (9.5 mm).

Where deluge systems are used, an adequate number of heat detectors shall be installed.

In mechanical induced-draft towers, heat detectors shall be located under the fan deck at the circumference of the fan opening and under the fan opening when necessary to comply with the following spacing requirements.

On towers having extended fan decks which enclose the distribution basin, detectors shall be located under the fan deck extension in accordance with standard indoor spacing rules for the type detectors used. Except where the fan deck extension Is. 16. ft. (4.9 m) or less, and the, dimension, is the length of the joist channel then only one, row of detectors centered on and at right angles to the, joist, channels is required. Spacing between detectors shall be in accordance with NFPA standards and local codes.

Heat barriers shall be installed under the extended fan deck to separate deluge systems when total number of systems exceeds the number for which water supply was designed. Heat barriers shall extend from the fan deck structure to the distribution basin dividers.

Where heat detectors are inaccessible during tower operation, an accessible test detector shall be provided for each circuit. In the case of pilot head operated systems, and inspector's test connection shall be installed on the pilot line and arranged to be accessible during tower operation.

Heat detector components exposed to corrosive vapours or liquids shall be protected by suitable materials of construction or suitable protective coatings applied by the equipment manufacturer.

Protection for Fan Drive Motor

A heat detector and water discharge outlet shall be provided over each fan drive motor when the motor is so located that it is not within the protected area of the tower.

Provision shall be made to interlock the fan motors with the fire protection system so that the cooling tower fan motors will be stopped in the cell(s) for which the system is actuated. Where the continued operation of the fans is vital to the process, a manual override switch may be provided to reactivate the fan when it is determined that there is no fire.

Exposure Protection

When any combustible exterior surfaces of a tower, including the fan deck, distribution basins, etc., are less than 100 ft. (30.5 m) from significant concentrations of combustibles such at structures, piled material, etc., the combustible exposed surfaces of the tower shall be protected by an automatic water spray system.

Systems for exterior protection shall be designed with the same attention and care as interior systems. Pipe sizing shall be based on hydraulic calculations. Water supply and discharge rate shall be based on a minimum 0.15 gpm/sq. ft. (6.11 L/min/m^2) for all surfaces being protected.

The design and installations shall comply with the applicable sections of NFPA 13, Standard for the Installation of Sprinkler Systems.

Piping, fittings and hangers, where exposed to atmosphere and inside cooling tower cells, shall be corrosion resistant or protected against corrosion by a suitable coating.

Approved discharge devices are made of nonferrous material and are corrosion resistant to normal atmospheres.

Some atmospheres require special coatings on the discharge devices.

Wax-type coatings shall not be used on devices without fusible elements.

Special care shall be taken in the handling and installation of wax-coated or similar sprinklers to avoid damaging the coating. Corrosion-resistant coatings shall not be applied to the sprinklers by anyone other than the manufacturer of the sprinklers, except that in all cases any damage to the protective coating occurring at the time of installation shall be repaired at once using only the coating of the manufacturer of the sprinkler in an approved manner so that no part of the sprinkler will be exposed after the installation has been completed. Otherwise, corrosion will attack the exposed metal and will in time creep under the wax coating.

Hydrants shall not be located closer than 40 ft. (12.2 m) to towers.

Towers with any combustible construction located on a building 50 ft. (15.3 m) or more in height shall be provided with Class III standpipe protection (as defined in NFPA 14) within 200 ft. (61.0 m) of all parts of the tower. Sufficient hose shall be provided to reach all parts of the tower. Provision shall be made for completely draining all exposed standpipe lines in winter. Home equipment at each standpipe hose connection on the roof shall be protected from the weather in a suitable cabinet or enclosure.

Areas around towers located on the ground shall be kept free of grass, weeds, brush, or combustible waste materials.

Smoking shall not be permitted on or adjacent to any cooling tower of combustible construction. Signs to this effect shall be posted and maintained and this regulation strictly enforced.

Forced and induced draft towers in continuous operation shall be checked frequently for excessive heating in motors and for excessive fan vibration.

At least semi-annually the fan assemblies including the motors and speed reducers shall be checked, both during operation and when shut down, for excessive wear or vibration, improper lubrication, corrosion, or other features that could result in failure and, where conditions require, corrective action shall be taken.

Where work on the tower requires welding or cutting, it shall be done in accordance with NFPA 51B and local codes.

Combustible cooling towers are particularly susceptible to ignition when they are shut down for repairs or other reasons and the wood becomes dried out. During these periods all automatic fire protection on the tower shall be operable, or if the tower is not so protected, special protection shall be provided until the tower is back in service.

Access to the top of water-cooling towers for firefighting and maintenance shall be provided by an approved stairway or ladder. Towers in excess of 120 ft. (36.6 m) in any dimension shall be provided with not less than two means of access remote from each other.

Motors, speed reduction units and drive shaft's shall be accessible for servicing and maintenance.

Provisions shall be made for lockout or tag out of electric motors when maintenance work is being performed in the area of fans.

After maintenance work is completed, all scaffolding, boards, temporary supports and other temporary materials shall be removed from the tower.

For proper maintenance of fire protection systems see NFPA 13A, Recommended Practice, for the Inspection, Testing and Maintenance of Sprinkler Systems.

43 Safety at the Work

The health and safety professional who expects to sell safety to management must get management involved in a total safety program. The emphasis must be on preventing office accidents as well as production accidents.

Planning for safety and health is an important part of every business decision, changes including purchasing, engineering, changes in work processes, and planning for emergencies.

If you have operations involving hazardous substances, procedures or processes, you must designate emergency response teams to be specifically trained and equipped to handle possible imminent hazards.

Toyotas' standard of performance is that which prevents accidents all together. Toyota is committed to provide working conditions that minimize hazards to health and safety. For example, Toyota has contracted an outside company to move all heavy boxes or materials for associates and insist that associates utilizes this service. The service is on the premises for a full day as part of our facilities group.

One reason office safety programs are not more widespread is that many people believe office injuries are minor. This is a serious mistake. Not only are medical and wage replacement benefits expensive for disabling accidents, but there are also hidden costs such as the loss of productivity.

Complacency — the attitude that office accidents do not amount to much — is one of the prime causes of office accidents. Office injuries may seem to be inconsequential because they lack dramatic impact. For example, the person who is injured falling backward from a chair seems to merit little sympathy or attention or it may seem amusing to some.

Industrial accidents are commonly considered to occur more frequently and with greater severity -- amputations, lost eyes, and broken bones, than office accidents. Office accidents may seem inconsequential because the often lack dramatization.

Training is one of the most important elements of any injury and illness prevention program. It allows employees to learn their jobs properly, brings new ideas and practices, and puts your program into action.

Accidents and injuries are not to be taken lightly, regardless of the nature of the accident or injury at least the National Safety Council and a few others don't think so.

For instance in California, Senate Bill 198 took effect in July. The new law mandates that employers form health and safety programs that include certain training, record keeping and management practices in the work place. The law requires that each associates' completion of the training be made a part of his/her employment history. It also requires that each associate be given not only general health and safety information but training on hazards specifically related to his/her job as well.

Safety committees have been formed at Toyota as well as making sure that all associates have access to information on potentially hazardous materials used or stored in the workplace. Researchers have found that new surroundings and length of service increase the chances of worker accidents and injuries. There is little difference between men and women office workers. An effective injury and illness prevention program requires proper job performance by everyone in the workplace.

Thorough investigation of all accidents and near misses will help identify causes and needed corrections, and can help determine why accidents occur, where they happen, and any accident trends. Such information is critical to preventing and controlling hazards and potential accidents.

The most common categories of major injuries are:

1. Falls from chairs, from slipping hazards, or from elevation.
2. Strains or other injuries related to overexertion.
3. Workers either being struck by objects (drawers, doors, filing cabinets) or striking them.
4. Workers being caught in or between machinery equipment.

Office accidents can be controlled by eliminating hazards or reducing exposure to them. Management can accomplish this goal most easily when the office is being planned, equipment purchased, or new procedures established.

Changing laws and Company policy also assign additional responsibility to every associate for compliance with safety rules and practices. Our company counts on each of us to observe the rules and to help ensure that others do as well. Offices should be laid out for efficiency, convenience and safety. The health and safety professional should see to it that hazards related to stairways, exits, aisles and passageways, doors, glass windows, lighting, ventilation, electrical wiring, equipment, and storage are properly controlled or eliminated.

Adequate lighting and proper office design can prevent glare or distracting reflections and help to reduce or eliminate eyestrain and visual fatigue. Ventilation should be carefully planned and installed by professionals to ensure that fumes, moulds, smoke, and other contaminants do not accumulate in office areas.

Management must protect office workers against electrical equipment hazards by using UL-listed equipment and by installing proper grounding and shielding equipment. Workers can be trained in the correct use of electrical machines and to place cords and wires so they do not become tripping hazards. All electrical supplies and equipment must be inspected regularly for shock hazards.

Moveable equipment must be properly attached to a desk or other surface, and heavy equipment should be braced against walls or columns. Floor surfaces should be slip resistant, and all carpets kept in good repair or replaced when frayed or warped.

Office chemical products represent another area of potential hazard. Workers must be trained in the safe use and storage of chemicals used copying, printing, and duplicating machines and for cleaning purposes. Where possible, nontoxic and non-flammable materials should be substituted for more hazardous chemicals.

The supervisor and safety professional are responsible for developing and enforcing safety procedures in office areas.

The following is a list of the Code of Safe Practices:

1. Close all file drawers immediately after accessing or returning contents.
2. Remove all extraneous objects that are in aisles or that block access to work areas.
3. Immediately clean-up spills on floors or report large spills/flooding conditions to Facilities or building management.
4. When lifting objects, use proper techniques. Recommended load limits: Men: 60 lbs. Women: 35 lbs.
5. Use step-ladders or stools to access elevated areas.
6. Use chairs on casters properly neither tipping them far forward or backward.
7. Open all doors cautiously. Do not run in office areas.
8. Report promptly malfunctioning electrical equipment and damaged cords and plugs to Facilities.
9. Smoking is allowed in designated areas only.
10. Open only one file drawer of cabinets at a time.
11. Use paper cutters with care keeping fingers and clothing out of the path of the cutting edge.
12. Close cover of xerography devices when copying to avoid looking directly into the camera light.
13. In case of fire, if in a small waste metal container, extinguish by "nesting" a like container on the top. Otherwise activate nearest fire alarm, notify supervisor and evacuate area according to plan. Know your evacuation route.
14. In case of earthquake, stand clear of glass walls and ceilings; take a position under a desk if shock waves are severe. Await instructions before leaving work area.

Workers should receive safety training from the first day on the job to prevent accidents and to spot hazardous conditions. They must know the safety procedures for objects, when working with office equipment, and while handling potentially hazardous objects or materials.

Employees should understand how to prevent fires in office work areas and what to do during a fire drill or outbreak of fire; 1) turn in the alarm, 2) alert fellow workers, and 3) use proper firefighting equipment.

Every company should draw up written emergency plans in the event of a natural or human-caused disaster. Every department should be assigned a specific route, and each worker should know his or her particular task.

The company should establish a formal safety program for office workers and establish training sessions for all employees. Training should focus on accident prevention, fire prevention, and emergency response, medical emergency response, ergonomics and stress-related disorders, and stress reduction.

The company needs to keep accurate records on all accidents and injuries. The records will show if the company is making progress in accident prevention. Standard report forms can be used to help pinpoint major problems so that management can design effective solutions.

Employers are reminded that the most cost-effective method is to implement and maintain a good safety plan, not wait to be cited or suffer in an accident.

Getting Help

1. Professional safety engineers are available as consultants.
2. Loss control experts are available from a company's insurer.
3. American Society of Safety Engineers
4. Veterans of Safety
5. National Safety Management Society
6. National Safety Council
7. Cal-OSHA's consulting organization, which provides free assistance.

If your company has an emergency guide, take a moment to review it so that you will be prepared in the event an emergency should occur. Note the type of warning signal which may be given for your building. Note the reporting area which has been assigned to your department.

Your safety and protection are the key elements of the company's emergency planning. You also play a role in helping to assure your own safety during an emergency.

44 Fire Safety

Introduction

With lives and property at stake, the best defense against fire is safety awareness. People trapped in a building on fire are quickly overcome if they do not act quickly. Some fatalities have occurred in as little as two minutes. The sad fact is that many fatalities need not have happened if people had known how to use the building built-in defenses in order to save themselves.

Experience and knowledge of issues in worker protection have led to a greater appreciation of how safety management is directly related to organisational effectiveness. To protect the well-being of employees is to protect the organizations' most valuable resource and to avoid the staggering costs and negative public image associated with safety neglect.

This discussion will be focused on three elements of fire safety: What you should know as an employer, what you should know as an employee, and proper evacuation procedures.

When a fire beings, grows bigger and then runs out of control and gets a firm grip on its surroundings, there is a drama in full measure. People die; corporate assets go up in smoke; livelihoods melt away in the heat—and somebody is eventually saddled with the blame. Unfortunately, you, as the employer, will get the heat.

The law requires that all employers must provide work and workplaces that are safe and healthful. Failure to do so can result in a threat to life or health of workers, and substantial monetary penalties.

Fire Prevention Plan

A written emergency plan spells out the basics and provides a framework for the future. Having a written program is very important so people can refer to it any time. It has to be

The best programs have one thing in common a custom fit—from the CEO to the first-line worker. Your plan has to wear well for your company at all its sites and in all departments. An emergency plan generally includes the following:

1. Fire department emergency telephone number 911 and other emergency response telephone numbers.
2. Evacuation and relocation plan for the building occupants.

3. Duties of the Fire Safety Director and other designated emergency personnel.
4. Building employee responsibilities in case of emergency, including individual assignment and responsibilities.
5. Procedures to identify and assist the non-ambulatory and physically disabled.
6. Basic information on fire systems or equipment found in the building, such as alarm devices, smoke detectors, fire extinguishers, sprinklers, and the like.

Training the Workforce

There is little point in having a knowledgeable management and supervisory body dedicated to preserving life and property from the ravages of fire unless the workforce knows about it. The men and women who handle the machines, work in the warehouses, operate the plant, clean up the depot and generally make things move need to let into the secret that a fire may ruin their job prospects.

Training does not necessarily mean knowing how to handle hoses and drive the fire engine, but it does mean that, at the very least, they can pick up the right sort of extinguisher and direct it at a mall fire with some prospect of success. Also that they know how to leave the building in a prompt but orderly manner when the fire alarm sounds.

Training is different from mere instruction. An employee can be instructed in how to use a fire extinguisher to the point where this person can recite the words of wisdom in his sleep. But unless he/she has picked one up, carried it to the fire ground, set it off and actually put out a stack of blazing wood or a tray of flaming liquid, then he/she is not properly trained.

Each company will have to decide upon the type of training which is required, the people who will benefit from it, the level at which it will be pitched, and the degree of receptivity which it needs to put across, the perceived weaknesses in existing performance which have to be corrected, and similar considerations.

Buying a "training package" from an outside organization is a possibility. This has a certain attraction because of its convenience, but this option rarely measures up to the specific needs of any one employer. Training packages are usually too general in their content, and employees need to be able to relate what they are shown and told to their own workplace, their own process, and those conditions which they recognize. In order to capture the interest of the employees in one's own organization, it is best, therefore, to structure the training by dealing with situations which are "domestic" and in which they have a vested interest.

Training will include a recognition of all factors within the process which have a bearing on fire risk, and a knowledge of procedures and safeguards which will eliminate or significantly reduce the risks which are identified. It will then deal with on-site equipment for putting out fires such as extinguishers, hose-reels, hydrants, automatic systems, and with much emphasis on the fact that knowledge and prompt action can save lives and jobs. Allied with this will be practical training for those employees who may be expected to use the equipment should the need arise.

It will also include the vital information on fire escape procedures, backed up by practical exercises involving evacuation drills and orderly mustering or personnel at assembly areas.

Built into this segment of training, of course, will be clear instructions and demonstration of how to raise the alarm and how to summon the Fire Brigade.

In addition to general training, there will also be particular training for members of the fire team, for appointed fire wardens, for the security staff, for the front office receptionist, and for all those people who may have specific jobs to do in the event of fire.

A fire can kill the person sitting next to you, your best friend in the office, your boss, the people you've worked with through thick and thin throughout the years, or even a family member working with you.

Here are some things that you should know about fire.

Classes of Fires

The National Fire Protection Association (NFPA) has categorized fire into four general classifications.

Class A Fires. This class of fires occurs in materials such as wood, cloth, and paper. Class A fires are extinguished by bringing the burning materials below the ignition temperature with the quenching and cooling effects of water. Under certain circumstances, these fires may be extinguished by the blanketing or smothering effects of dry chemical and carbon dioxide fire extinguishers.

Class B Fires. Class B fires are those that occur in the vapour-air mixture over the surface of flammable liquids such as oil, greases, alcohol, kerosene, gasoline, etc. Class fires are most successfully extinguished by limiting of the air which supports combustion. Fire extinguishers of dry chemical, carbon dioxide, foam, halogenated hydrocarbon agents, and fog streams of water are recommended for this type of fire.

Class C Fires. Those which occur in electrical equipment where nonconducting extinguishing agents must be used. Dry chemical, carbon dioxide, and vaporizing liquids are suitable. Because foam, water (except as a spray), and water-type extinguishing agents conduct electricity, their use can kill or injure the person operating the extinguisher, and severe damage to electrical equipment can result.

Class D Fires. These occur in combustible metals such as magnesium, titanium, zirconium and sodium. Specialized techniques, extinguishing agents and extinguishing equipment have been developed to control and extinguish fires of this type. Normal extinguishing agents generally should not be used on metal fires as there is danger in most cases of increasing the intensity of the fire because of a chemical reaction between some extinguishing agents and the burning metal.

The Fire Triangle

To begin and survive, a fire needs three components — heat, oxygen, and a fuel. Just two of these are not enough. And if any one of them is not present, there won't be a fire. By the same token, if there is a fire, remove one of the components and fire will go out.

Once a fire has started, putting it out is a matter of removing one of the limbs of our triangle.

Fuel can be removed from the scene in a number of ways, depending upon its nature, the quantity involved, the scale of the fire and the availability of access, and the machinery of removal. One method may simply involve taking wooden or clothing materials away from the scene. In all cases, the aim is the same -- to starve the fire of the fuel it needs to survive. Obviously, there are many occasions when the fuel cannot be removed. Once the fabric of a building is ablaze, the timber, papers, furniture, fabrics, plastics, and flammable chemicals which it contains will have to stay put until they are burned up or until the fire is extinguished.

Heat is removed by the effective application of water and, again, it depends upon the scale of the fire and the availability of the water. A bucket of water, or the jet from a two-gallon water extinguisher, will serve the purpose of taking the heat out of a fire in a waste-paper bin. When it comes to a major outbreak in a factory building, or any other large fire, the scale of attack can range right up to many thousands of gallons of water per minute. At either end of the scale, the aim is the same — to cool the fire to a point below that of the ignition temperature of whatever fuel is burning.

Oxygen can be removed from the equation by the exclusion of air from the surface of the burning object, so that the vapour which arises cannot mix with the oxygen in the air — a smothering process. Removal of the oxygen can be effected in several ways; for instance by putting the lid over a pan of blazing fat; by laying a blanket of chemical foam over the surface of a flammable liquid; and by laying a cloud of "dry powder" over an oil or petroleum fire.

The chemistry of fire, then is simple but essential. Fire needs heat, fuel and oxygen; to stop fire from happening, or to put it out, it must be cooled, starved or smothered.

Fire Extinguishers

Since the overwhelming majority of major fires have very small beginnings, it follows that we must be prepared to deal very promptly with minor outbreaks at the place of work or anywhere else. The essential thing is to have the right type of equipment in the right place, and to have people on hand who know exactly what to do at that crucial moment when smoke and flames begin to curl.

Each extinguisher has a label. The label indicates what class of fire it can be used on effectively. Some common types of extinguishers are:

1. Water-based. The most common type of water extinguisher is the pressurized type. It comes in a standard 2½ gallon tank. This size is dictated by the need for a sufficient amount of water to put out a small fire, and still have a container that is not too heavy for most people to lift. The 2½ gallon size weighs around 35 pounds. The water extinguisher will last for one minute of continuous operation. It will squirt about 35 to 40 feet, and is pressurized to 110 PSI. Water at 110 PSI through a 1/8th inch tip will Have enough force to scatter light weight materials. If it is pointed at a waste paper basket at close range, the force of the water can overturn the container and cause the fire to spread. To get in close with a water extinguisher of this type, it is necessary to break the stream of water into a spray by bouncing it off the tip of your finger with a garden hose. The water extinguishes the fire by absorbing heat and cooling the fuel below its ignition temperature. Water extinguishers are to be used on Class A fires only.

2. Dry Chemical. There are two categories of dry chemical extinguishers sold. They are commonly designated BC type and ABC type by the kinds of fires they will work on. The BC dry chemical extinguisher commonly contains bicarbonate of soda or potassium bicarbonate (Purple-K). These units will extinguish B and C fires, but are not particularly effective against common combustible fires. The mono-ammonium phosphate commonly found in ABC dry chemical extinguishers will work on all three types of fire.

All dry chemical extinguishers will perform in about the same fashion when they are being used. The 2½ pound size will empty in about 8 seconds of continuous use. The larger units empty in from 10 to 20 seconds.

The dry chemical is very fine grained, almost flour-like in its consistency. This powdery quality forms a cloud as it is expelled that will leave a dusty residue on all surfaces in the area which will need to be cleaned off. BC and ABC chemicals are not compatible for use on the same fire, at the same time. One may be used, and then the other, but not simultaneously because the two chemicals tend to interact and each tends to neutralize the action of the other. Both types of dry chemical agents use a flame inhibiting mechanism which operates on a chemical level. By almost instantaneous flame suppression. The fire does not have to be physically buried in the chemical to go out. All that is necessary is for a cloud of the material to be present in the area where the chain reaction of the fire is taking place.

3. Carbon Dioxide (CO_2). The advantage of the carbon dioxide extinguisher is that it is clean. The carbon dioxide evaporates directly into the atmosphere and leaves no residue behind. This makes it fairly popular in kitchens and other places concerned about mess and contamination. CO_2 extinguishers are only good on B and C fires. They are only ½ to ⅓ as effective, pound for pound, as a BC dry chemical extinguisher. Their range is between 3 and 8 feet which forces the person using them to be relatively close to the fire to be effective. Most units are noisy in their operation. CO_2 puts out fires by smothering them. The relatively inert character of the gas and its heavier than air quality form a gaseous lid on the fire separating it from the atmospheric oxygen. Vessels containing carbon dioxide are self pressurising to 1800 PSI. This high pressure requires that the walls of the container be strong and thick, and therefore heavy.

A carbon dioxide extinguisher will handle a small grease fire on a stove top. It will work well on electrical equipment fires, because it doesn't leave a heavy coating of powder.

CO_2 doesn't work well on Class A fires.

4. Metal Extinguishing Agents. The agent used will vary for different metals, so it is essential to have the proper agent for the specific metal which might catch on fire. The main characteristic of these agents are their resistance to being broken down by intense heat. Because heat does not destroy it, it can be used as both a smothering, and a cooling agent. The agent must be applied over the entire surface of the burning metal to form a sufficient blanket to keep the oxygen away from it. Dry sand, graphite, and metal-x-powder are some of the special agents used on some metal fires.

How to Operate Fire Extinguishers

An acronym, PASS, is used to remind people on what to do when operating a fire extinguisher.

It stands for Pull, Aim, Squeeze, and Sweep.

1. Pull the pin at the top of the extinguisher. It is between the handle and the operating lever, held in by a plastic or fine wire seal. The seal prevents an accidental discharge. To pull out the pin, it is necessary to break the seal.
2. Aim at the base of the fire. To aim the extinguisher, it may be necessary to physically tip its entire body so that the nozzle points at the base of the fire.
3. Squeeze the trigger/lever down against the lower handle by squeezing on it or by leaning on it with the heel of your hand while the base of the extinguisher rests on the floor.
4. Sweep the fire side to side to put out the flames.

Fire Prevention Tips

Here are some fire safety reminders:

1. Know the exact location of fire alarms and how to operate them.
2. Be familiar with the location of all fire exits. Posted floor plans specify the fire exits.
3. Learn how to operate a fire extinguisher.
4. Be alert for any defects in equipment or machinery which could result in accidents and injury. Call for repair or replacement for frayed electrical cords, damaged cables, split hoses, worn-out hand tools, etc. Do not overload electrical sockets.
5. Be familiar with telephone numbers to call in case of fire.
6. Store flammable liquids properly.
7. Clean up waste immediately. Dispose trash in approved containers.
8. Keep aisles free of obstacles.
9. Be familiar with the contents of your fire prevention plan, especially emergency procedures.

Evacuation

Some people feel the best thing to do in a fire is to run out of the building, to evacuate immediately. This is not always the case. Different situations will dictate a variety of responses that might be better. It may not be practical or even possible to leave the building quickly. Sometimes it may be wise to go up or down in a building. Other times, it is best to stay put.

In a high-rise building, jumping is not recommended. People will not usually survive the 20 to 25 miles per impact they experience when jumping from the third floor. However, if you are going to exit from the second floor, you may be able to go out of the window on your stomach, feet first. Hang down by hanging on to the window sill and just let go. Most of the time, your feet will only be about six feet from the ground. Do remember that's only from a two story building. Look before you leap!

Procedures

Upon the Discovery of Fire

1. Keep Calm.

2. Get the people out. Do not attempt to fight the fire at this time unless it is the only way that you can save a life If you get hurt, you may not be able to report the fire.
3. Close the door to confine the fire once the people are out. A closed door will hold the fire and buy you time. If the door is closed, the fire may be confined to a single room.
4. Give the alarm. You must notify the people in the building and the Fire Department. Information to be transmitted: 1) Location of the fire (address, nearest cross street, portion of the building, etc.); 2) Telephone number; 3) Type of fire; 4) If known, explain what is needed, i.e., paramedics, police, or security.
5. Fight the fire. If it is safe to do so, return and extinguish the fire using first aid fire equipment.
6. Evacuate. If the fire continues despite your efforts, get out! Close the door. Now is the time to start considering the necessity of moving people away from the fire area.

When Evacuation is Required

1. Move quickly, but do not run.
2. Go to the nearest designated stairwell or exit. Do not use elevators. The posted floor plan gives you the best exit route.
3. Remove high heels to prevent injuries (carry them with you).
4. Give assistance to the disabled.
5. Go to the designated meeting place.
6. Resist any temptation to go back into a burning building. If someone is trapped inside, firefighters equipped and trained for rescue have the best chance of helping the trapped person.
7. Pray. People may have different religions, but there is only one God.

Fire does not occur everyday. But when it does happen, you should be prepared. The key is simply knowing what to do and what not to do in case of fire.

A colourful picture of safety tips posted on the wall may paint a thousand words, but knowing how to act on them is worth a million.

45 Industrial Explosion Prevention

Explosions have touched many industries: chemical, petrochemical, food processing, brewing, pharmaceutical, and more. Some building demolition or quarry blasting, for example are planned and controlled. Many, however, are not. Accidental industrial explosions harm personnel, damage property, and interrupt business. Anyone who handles flammable substances or combustible powders is vulnerable to the consequences of an explosion.

Losses from industrial explosions total millions of dollars each year. Business losses may equal or exceed property losses. Understanding the elements that contribute to such explosions and how to control them can reduce their frequency and severity.

What is an Explosion?

An explosion is the result of the rapid release of internal energy. This energy can develop from a chemical reaction with or without combustion, or it may have a physical cause. Explosions caused by chemical reactions fall into two categories: deflagration and detonation.

A Deflagration is a reaction that produces heat and propagates to the unreacted material at a speed less than sound. A Detonation is reaction that also produces heat and proceeds to the unreacted substance faster than the speed of sound. Shock waves and excessive pressure accompany the reactions. It is important to remember that a deflagration can become a detonation under certain conditions, such as a long run of pipe.

For a combustion to occur, a fuel and an oxidizer must mix together in appropriate quantities and be exposed to an ignition source of sufficient strength. For an explosion to occur, the reaction must be confined or partially confined. Because large quantities of gas are generated at high temperatures, the pressure inside the confined area also increases.

The maximum pressure developed during an explosion can be measured by American Society for Testing and Materials Standard (ASTM) E1226 "Standard Test Method for Pressure and Rate of Pressure Rise for Combustible Dusts." The test, conducted in a 20-litre laboratory explosion chamber, is used to quantify explosion severity.

Maximum pressure does not depend on vessel volume.

Therefore, the maximum pressure generated in a 20-litre vessel is also the maximum pressure generated by explosions in larger, industrial-sized vessels. Because the test method seeks the maximum explosion pressure over a range of fuel concentrations, the pressures stated normally provide an adequate safety margin and can be sued as the basis for the design of explosion protection devices and methods.

The rate of pressure rise indicates how fast the pressure increases within the confining vessel. For example, a sugar dust explosion could cause the pressure in a vessel to reach 8.5 barg (barometric pressure, gauge) within milliseconds after ignition. Most industrial vessels are not designed to withstand such pressures and require some sort of explosion protection. Explosion protection methods include prevention, containment, venting, suppression, and isolation.

Prevention and Containment

Explosion prevention is intended to minimize the probability of accidental explosions by keeping a combustion reaction from occurring. One way to eliminate sources of ignition energy. Common sources sufficient to cause explosions in industry include flames, hot surfaces, spontaneous heating, welding or cutting operations, friction heating or sparks, impact sparks, electrical sparks, and electrostatic discharge sparks.

Explosion can also be prevented by keeping the fuel or oxidant concentration outside the combustible range. Each combustible gas/air mixture has a lower and an upper flammability limit (LFL and UFL). The LFL is the minimum concentration of combustible gas in air below which a flame will not propagate in the presence of an ignition source.

The UFL is the maximum concentration of combustible gas above which a flame will not propagate in the presence of an ignition source. Gas explosions can be prevented by maintaining the gas concentration below the LFL or above the UFL.

Dust concentrations must be maintained below the minimum explosive concentration.

It may not be economically or physically viable to design a vessel to the standards for containing an explosion. Both the primary vessel and its appendages would need to be designed for containment. Because an explosion can propagate with or against the flow of the process, connecting vessels must also have sufficient strength for containment.

Containing an explosion should always be considered, but it often is practical only in small operations.

Venting

If prevention methods are impractical, or if the risk is great enough to warrant additional protection, other methods can be used to supplement prevention techniques.

National Fire Protection Association Standard 68 "Venting of Deflagrations" contains guidelines for designing and using vents for explosion protection. The pressure generated by a deflagration causes the weakest part of a vessel or enclosure to rupture first. A vent designed and placed on the vessel so that it is the first area to rupture minimizes the structural and mechanical damage caused by combustion pressures. Venting also directs the flow of combustion gases and unburned material away from personnel to the outside.

The reduced pressure P red, is the maximum pressure developed during the venting of a deflagration and is usually much lower than the maximum explosion pressure, Pex.

Venting does not extinguish the combustion reaction; it only relieves pressure and decreases damage in the primary vessel. When a vent opens, it exhausts both hot gases and unburned material. The unburned material will likely continue to burn, creating a secondary hazard. If connecting piping contains combustible materials, a flame may propagate into other places. Venting is not a suitable technique to apply when environmentally toxic materials are involved.

Suppression

Explosion suppression devices detect an explosion in its early stages and introduce a suppressant that prevents the combustion reaction from continuing. Some common suppressants are water, halogenated hydrocarbons, ammonium phosphate, sodium bicarbonate, and sodium chloride.

Suppressant agents prevent the combustion reaction from propagating by:

1. Absorbing energy produced by the combustion reaction and essentially cooling down the area in combustion;
2. Inhibiting the critical chain reactions propagated by the combustion reaction;
3. Lowering the combustible concentration below its LFL.

All three activities contribute to the effectiveness of a suppressant.

Speed is critical in a suppression system. The agent must be delivered a few milliseconds after the explosion is detected. Suppressant systems must detect an explosion in its early stages, then quickly deliver sufficient quantities of suppressant to keep the combustion pressure below the safe working pressure of the vessel.

Two common types of detectors are optical and pressure devices. Optical detectors may operate on infrared or ultraviolet principles. Optical detectors may operate detectors may operate on infrared of ultraviolet principles. They are used to monitor a vessel or an area and trigger appropriate control measures when a combustion flame is sensed. Optical detection systems must be carefully designed to prevent false alarms caused by dust and dirt, optical detectors are not typically used to protect dust systems. Pressure detectors are used to protect both gaseous and dust systems. They are used in conjunction with a control panel to provide an alarm or signal if the pressure exceed a present level. The present point takes into account normal pressure perturbations, yet the detector is set to sense an explosion in its incipient stage.

When an incipient explosion is detected, a signal from the control panel initiates the release of the agent. The cloud of suppressant interacts with the growing fireball that results from the explosion and prevents the propagation of the combustion reaction.

Suppressants are stored in a pressurized cylinder designed to deliver the agent rapidly. An electrical signal from the control panel opens a rupture disk at the end of the suppressant cylinder. The pressurized agent is release pneumatically into the combustion zone. The agent often is released through a nozzle to ensure total coverage of the area.

Explosion suppression systems limit pressures caused by explosion and prevent fire or explosion from propagating into other vessels. Suppression does not require the release of the vessel contents or combustion products into the surrounding atmosphere. Suppression systems are more expensive than venting and require more maintenance.

Isolation

When several interconnected vessels handle combustible materials, they may each need to be protected or isolated from each other to prevent explosion propagation. The effects of an explosion are greater in subsequent vessels.

In interconnected vessels, the first explosion increases the pressure in the entire system and secondary explosion occurs at higher initial pressures, resulting in more severe damage. This phenomenon is called pressure piling. It can be minimized with an explosion isolating system. There are two types: mechanical and extinguishing.

Mechanical isolating devices include the explosion isolating valve and the backflash interrupter.

An explosion isolating valve closes rapidly if an explosion occurs. A detector and a control panel are integral to the system. As with the suppression system, when the control panel senses an explosion, it triggers the rupture disk that opens and release pressurised nitrogen. The release of a nitrogen pushes a piston and closes the valve, preventing the propagation of the deflagration wave into other vessels.

A backflash interrupter (BFI) is a passive mechanical isolation device currently being developed. It requires no detection or control system.. If an explosion occurs, pressure from the explosion opens the dome on the BFI, interrupting the propagation of the deflagration wave and releasing the pressure into an open area.

Extinguishing isolation systems are relatively new, and their benefits are still being discovered. They consist of a detector, control system, and release container filled with suppressant. They are similar to explosion suppression systems and explosion isolation valve systems. When pressure from an explosion is detected, the control panel initiates the protection sequence.

Total Protection

Individual explosion protection methods can be used together to provide total protection. Small hazardous processes may be protected by containment. Suppression systems may be more suitable for larger ones. The detection and control system of an explosion protection system can interface with other process control functions or fire protection systems to limit damage.

Because explosion protection is a developing field, accurate and current data must be used when applying explosion protection systems. A system may need to be tested in an experimental manner before it is applied in the plant. Combustible materials will always be of importance of industry, and the hazards associated with handling them must be minimized in a practical and economical way.

46 OSHA

Introduction

OSHA is an acronym that stands for the Occupational Safety and Health Act. This Act was signed into law (Public Law Number 91-956, United States Code (USC) 651-678) on 29 December, 1970, by then President Richard Nixon (National Safety Council, 1992). As its name implies, this Act deals primarily with the safety and health of employees on the job. OSHA encompasses workers on both the federal state level. States may be allowed to have differing programs. The only "catch" is that they can meet or exceed federal standards and regulatory factors. They may not fall below established federal guidelines.

A brief history of the rules concerning the laws affecting the safety and health of the workers of this country shows that the states were pretty much left in charge of this regard. Sometimes referred to as the "Williams-Steiger's Act," the important feature of this piece of legislation is that it addresses concerns "beyond the present workplace, and considers long-term health hazards in the working environment of the future (National Safety Council, 1992). We can all see how this is an important factor for all employees, as with ever-increasing technology often comes with differing kinds of long-term health hazards, e.g. radioactive materials.

Prior to this Act, the two main laws that addressed these concerns included the Walsh-Healey Public Contracts Act, and the Longshoremen's and Harbor Worker's Compensation Act. Some others that followed included the Service Contract Act of 1965, and the Contract Workers and Safety Standards Act (Construction Safety Act).

OSHA also stands for the Occupational Safety and Health Administration. Created in 1971, the Administration has four main areas of authority. These are: "to promulgate, modify, and revoke safety and health standards; to conduct inspections and investigations and to issue citations, including proposed penalties; to require employers to keep records of safety and health data; to petition the courts to restrain imminent danger situations; and to approve or reject state plans for programs under the Act (National Safety Council, 1992).

The Administration was the product of the Department of Labour (DOL). However, the fines or citations issued by OSHA is not without a remedy for those who feel that their situation is not being handled in a fair manner. There exists a review commission called the Occupational Safety and Health Review Commission (OSHRC). To preclude any bias toward the government's favour or against the contractor or employer involved, the three members are appointed by the president and confirmed by the Senate. However, they, (the Commission) are a separate entity to themselves and do not fall under the auspices of the Department of Labour. The commission has the last say on these matters. As with any litigation, however, appeals to this may be debated in the courts.

Applicability of the Act is Broad and Encompassing

Exclusions include federal and state workers. This, however, is due to the fact that the federal and state organizations are required to have OSHA minimum standards, so, in effect, they are covered. Beyond the exclusions, though, OSHA "applies to every employer who has one or more employees and who is engaged in a business involving interstate commerce (National Safety Council, 1992)."

The Act holds employers responsible for ensuring a hazard- free environment in which the employee is to work. It also holds employers responsible for abiding by the rules stipulated by the Act.

Among the rights given to employers by this Act, it is the right to know the reason for the inspection, if one is given. Also, OSHA inspectors are required to carry proper OSHA identification. They must wait for confirmation if the employer has doubt as to his or her authenticity. The employer need only to call OSHA local region to verify. Employers can also ask for guidance from the Administration as well as a variance if materials are not readily procurable within the required time specified.

Moreover, employers can contest the findings of OSHA regional authorities, as mentioned earlier, with the Review Commission.

Why are these issues concerning this Administration important and relate to the welfare of the people? Because the costs of accidents, both on the small, personal scale, and on the large aggregate are so high. Examples include "productive time lost by an injured employee, productive time lost by employees and supervisors attending the accident victim, and clean up and start-up of operations interrupted by the accident. Others include the time that it takes to hire or retrain other individuals to replace the injured worker until his/her return to work; the time and cost for repair or replacement of any damaged equipment or materials; the cost of continuing all or part of the employee's wages in addition to compensation; the reduced morale among employees, and, perhaps, the lower efficiency; increased workers' compensation insurance rates; and the cost of completing the paperwork that is generated by the incident" (Cal/OSHA, 1993).

All these reasons are drivers behind California's OSHA regional directives. To help in getting an organization started in the right direction towards regulatory compliance, Cal/OSHA directs a workplace assessment be done as part of an evaluation of an employer's injury and illness prevention program. Factors can include: accident, injury, or illness data, rates of employee turnover or absenteeism, company policy statement, guidelines for proper work practices or procedures, records or training programs, and compliance with the requirements of California Right to Know Law and Hazard Communications Standard (Cal/OSHA, 1993).

Perhaps one of the most important parts of the assessment is that of categorizing and documentation of employees' capabilities. This helps to facilitate the proper placement of employees into their respective areas matched by their abilities. "By applying occupational health principles to the workplace, all employees, including the severely disabled, are placed in jobs according to their physical capacities, mental abilities, and emotional makeup. This approach also ensures continuing medical care and rehabilitation of occupationally ill and injured workers (National Safety Council, 1992).

After this assessment is done, Cal/OSHA recommends that care should be given in making sure that the data gathered from the assessment accurately corresponds and complies with OSHA standards. If not, remedial action to correct such should be applied as soon as possible.

Finally, they direct the formation of an action plan and the steps to taking the operation into a state of putting the plan into action and practice. It does no good to assess the workplace environment without following through with a decided course of action.

For those employers not so inclined to follow the established rules set forth by the Act, and enforced by the Administration, the Administration can take recourse against such by levying fines and civil penalties to negligent employers. The differentiation among these factors and contingent upon several issues. The first that is assessed is the types of violations(s). OSHA sees two types: serious, or other than serious. Once this has been decided, the next step is the determination of the "gravity of the violation." Of these, OSHA sees four types: high severity, medium severity, low severity, and minimal severity (National Safety Council, 1992). These penalties are assessed in the aggregate.

However, if the Administration determines the violation(s) to be egregious of flagrant, OSHA will levy a separate penalty for each violation. Again, any employer has the right to contest and OSHA action.

With regards to the relationship between the federal government and the jurisdiction of the states as concerns occupational safety and health, OSHA "encourages the states to assume the fullest responsibility for administering and enforcing their own occupational safety and health laws" (National Safety Council, 1992). In as much as this true, as mentioned earlier, the state may not have criterion that is below OSHA standards. The criterion must meet or can exceed the standards set forth. Moreover, the state can opt not to address or cover a certain area in their state plan.

However, in any area(s) that the state does not cover, in that area the state "surrenders its jurisdiction for safety and health programs."

As OSHA deals with safety and health, it is easily understandable, in today's world of high-technology and large company concerns, that there is a direct and serious correlation between occupational safety and health and the environment. The Comprehensive Environmental Response, Compensation and Liability Act (CERCLA), and the (RCRA) which stands for the Resource Conservation and Recovery Act, were important acts of legislation with regards to occupational safety and the environment.

OSHA has gone so far as to have signed a memorandum of understanding with the Environmental Protection Agency (EPA). One of the areas of focus is on the petrochemical industry. These environmental laws also "impose requirements for the use of protective equipment and procedures, and for extensive reporting to federal, state, and local government agencies" (National Safety Council, 1992). In addition, "they also affect the design and operation of many manufacturing processes and waste disposal systems." (ibid).

CERCLA is also known as the "Superfund," and was created in 1980. It was created because of the realisation of the existence of uncontrolled, dangerous disposal sites of toxic chemicals. These sites are a danger to those both working on the sites and those living nearby. While there have been some problems in funding the "Superfund," in the past, the effort is a tool that is used by the government to address environmental cleanup issues. It has been of help in cleaning up the lakes and managing landfills around the nation.

The bottom line as regards OSHA, is that it is both an important piece of legislation and an administration whose service is needed both in the current work environment(s) of the country, and will have an ever-growing importance in the working environments of the future. With today's technology, things can only get more complex, and we need a system of safety engineering and regulatory controls that OSHA can provide.

47 Fire Safety in the Home

The average fire is far more devastating and life threatening than the average burglary. Each year about 7,800 people die in home fires. Monetary damages average well over two billion dollars annually. No one wants to think about the possibility of a fire happening, but we all agree it's much better to be safe than sorry.

Even with an early warning from smoke detectors, escaping a fire can be difficult or impossible. Fires can spread very rapidly, blocking exits and creating dangerous smoky conditions. It is important that all family members know what to do when the going gets tough. Try to remain calm and do not panic. Focus all of your attention on how to get out safely. Get out and stay out! Go as quickly as possible leaving all possessions behind. And never go back in, you may not get back out. If you hear the detector alarm, smell smoke, or suspect a fire, feel the door. If it is hot, try your other exit; if not, slowly open the door but be prepared to close it if smoke of flames rush in. Don't try to go through fire unless you have no other course of action available to you. Smoke is your enemy! Even a few breaths of smoke can choke and kill you. If you become trapped in smoke, crawl low, keeping your head down. Smoke and heat rise, so cleaner air is near the floor. If smoke, heat, or flames block your escape, stay in the room with the door closed. Open a window at the opt and bottom as long as no smoke is entering the room. Do not break the window.

Keeping low, put a wet cloth over your nose and wait at the window, signalling with a bright sheet or flashlight. If there is a phone in the room, call the fire department and tell them where you are and that you are trapped.

Planning Your Escape

If a fire occurs, there's no time for planning. Everyone in the family needs to know in advance the best ways to evacuate the home, so they can move quickly — and without panic — when the alarm sounds. That's why planning and practicing escape plans is so vital. If you move make a new fire escape plan immediately.

The first step is to draw a complete floor plan of your home. Then mark all possible escape routes. Colour-code primary exits green; secondary exits yellow; dead ends red. Remember there should be at least two ways out of every room in case smoke or flames block your primary exit. Make sure you know how to open windows that are designated as escape routes and be sure that they open easily. Decide on a meeting place outside where all family members assemble after exiting the home. This way, you can make sure that everyone is out.

Post your fire-escape floor in convenient, central areas of the house where guests and baby-sitters can see it.

Heating Sources

Home heating sources such as fireplaces, wood stoves, portable heaters, and furnaces can be fire hazards if improperly operated and maintained.

Furnace

Have your furnace checked every year? Be sure all furnace automatic controls and emergency shut-offs are in good condition. Leave furnace work to experts. If the pilot light on a gas furnace goes out, have a qualified specialist turn it on for you. Don't attempt repairs unless you are qualified. Have the repairman check the wall and ceiling near the furniture and flue. If they are hot, additional insulation or clearance may be needed. Check the flue pipes. Are they well supported? Free of holes and clean? Is the chimney solid? No cracks or loose bricks? Uninsulated flue pipes should be at least 18 inches from any combustibles (6 inches for gas furnace). All unused flue openings should be sealed with solid masonry. Keep trash and combustibles away from the heating systems. Don't store hot ashes in the home; take them outside immediately. Keep the basement door closed and make sure the door fits tight.

Fireplaces

Keep a metal screen in front of your fireplaces; flying embers can start fires. Use a spark screen or heat resistant glass doors. Clear the area of all combustibles, especially rugs. Store firewood at least 3 feet from the fireplace and burn only dry firewood. Don't use excessive amounts of paper to build roaring fires in fireplaces, and never use flammable liquids to start the fire. It is possible to ignite soot in the chimney by overbuilding the fire.

Never burn charcoal in your fireplace, or in a charcoal broiler or hibachi unit inside. Burning charcoal gives off deadly amounts of carbon monoxide. Don't burn plastic or garbage in the fireplace; plastic can emit toxic fumes and garbage can burn out of control. Never burn painted or pressure treated wood; the wood can emit toxic fumes. Follow the direction on the package if you use man made logs. Never break a man-made log apart to quicken the fire.

Smoke Detector

Most fatal residential fires happen between 8 pm and 8 am when occupants are more likely to be asleep. Smoke detectors help alert you to a fire before it spreads to you. That way, you've got a chance to get out while there's still time.

Smoke detectors are easy to install and maintain. The national Fire Protection Association has good news and bad news about smoke detectors. The good news is that smoke detector use in the United States rose in 1987 to four-fifths of all homes. The bad news is that in roughly one-third of these homes, the detectors don't work. This means that more than one-fourth of all American homes have non-operational smoke detectors. Power-source problems, especially dead and missing batteries, are the main reason for non-working smoke detectors. Dead batteries probably reflect the lack of regular testing and maintenance, while missing batteries may reflect a wider range of problems, including frustration over nuisance alarms. To combat this problem, the National Fire Protection Association advised that you change the batteries in your smoke detectors each year on the daylight to standard time change.

Fire Escape Ladders

Homes with bedrooms more than one story off the ground should be equipped with fire escape ladders so that occupants of those rooms can use the window as an escape route in case the stairwell is blocked. The best type of fire escape ladder is the portable type which installs easily, in inexpensive and inconspicuous. Such a ladder should be made of chain or steel cable and fitted with spacers or "standoffs" to keep the rungs away from the wall for toe and handholds. One type of ladder requires no installation. It is kept in a box out of the way, then quickly taken out and hung from the window frame when needed. Other types are anchored permanently near the window.

To use the ladder drop the rungs out of the window and hook the frame over the sill. Climb sideways out the window, straddling the sill so that you can see the ladder beneath you and place your foot on a convenient rung. Have only one person climb down the fire ladder at a time. The first person down should hold the bottom of it steady for the next person. To carry a small child, have him wrap his arms and legs securely around you, then climb down the fire ladder face to face with him. Place an infant in an infant carrier and climb down the fire ladder the same way. Make sure that you practice using your escape ladder so that you can move quickly in case of an emergency.

Home Sprinkler Systems

Sprinkler systems could prevent thousands of fire tragedies, especially in one and two-family dwellings, where four-fifths of all fire deaths and more than $4 billion in property damage occurred in 1988. With a goal to save more lives, as well as to operate within tightened fire-protection budgets, hundreds of U.S. communities have passed residential fire sprinkler laws. While only about twenty-eight cities and counties in the United States require sprinklers in one and two-family homes, laws requiring them in multi-family dwellings are much more common. Contributing to the increasing popularity of these ordinances is the fact that modern fire sprinkler systems are more economical than in the past.

Until recently only iron pipe was used to carry water from the main supply to the sprinkler heads. Now less-expensive copper, lightweight steel, and plastic CPVC or polybutylene pipe has been approved for sprinkler systems. Residential sprinklers use water from the household supply, another factor which makes them more economical than commercial systems, which often require dedicated water supplies. Designed with small water supplies in mind, residential sprinklers are more sensitive to heat that standards sprinklers, and are triggered in the early stages of a fire when the fire can be controlled with less water. When the temperature reaches 165°F., only those sprinkler heads closest to the fire open to spray water on it.

Sprinkler heads designed for residences are smaller and more attractive than commercial heads, also adding to their popularity. Residential sprinklers can be ceiling or wall mounted, and are available in flush and recess-mounted styles. Fire sprinkler systems can be installed in newly constructed buildings or existing homes. A sprinkler system for a typical single-family home with a good municipal water supply costs the home owner between $1,500 and $42,500, according to the National Fire Sprinkler Association. For a home under construction, it usually takes two to three installers a day to a day and a half to put in the sprinkler piping, and another day for the fire inspector to test the system. Whether installed in new or existing homes, sprinklers often mean homeowner's insurance discounts. Carriers reduce premiums 10 to 20% when a home has an automatic fire sprinkler system.

Fire Ratings

Common coverings for walls and ceilings are grouped into four classes by "flame-spread ratings." Established by the National Bureaus of Standards and national testing laboratories, these ratings are based on a comparison of the materials' burning speeds with those of asbestos (specified as Zero) and dry red oak (100). Class A and B materials are recommended by fire-prevention experts for halls, stairways, kitchens, and utility rooms. Except in these locations, Class C materials may be used to cover small areas, but not an entire room. Class D materials are unsuitable for home use; they do not meet minimum standards of the federal government. Manufacturers of building materials mark many of their products, including acmustical ceiling tiles and wall paneling, with flame-spread ratings.

48 Explosive Dust

Reasonable care shall be exercised to prevent personal injuries or property damage arising from fire or explosion caused by flammable dusts.

Any person, firm, or corporation intending to install a dust collection system, or remodel or enlarge an existing system, shall first submit plans in triplicate to the Chief for approval prior to installation. Approval by the Chief shall be tentative only, and subject to final approval on inspection of premises and systems when in operation. It shall be unlawful for any person, firm or corporation to install, remodel, enlarge, or use such a system without the approval of the Chief.

Housekeeping and Maintenance

1. No person shall permit the accumulation of any flammable or explosive dust in quantities sufficient to create a fire or explosion hazard on floors, walls, ledges, or other exposed building structural members, piing, conduit, electrical equipment, or on the roof of any building on which dust may settle.
2. The use of compressed air or other means to blow dust from ledges, walls, and other areas which may create a dust explosion hazard shall not be permitted unless all machinery in the area has been shut down and all sources of ignition removed.
3. The removal and collection of small quantities of loose dust may be accomplished with approved industrial vacuum cleaners.
4. Metallic dust shall be stored in tightly covered metal containers in a cool dry place.

Dust Collection Systems for Rooms or Buildings

Every plant, factory, workshop, or other premises in which dust is produced, or is a by-product, in such quantities as to create a hazard, shall be equipped with an approved pneumatic or mechanical dust collection system designed, maintained, and operated to keep all hazardous accumulations of dust from such premises, including all enclosed an inaccessible areas, and spaces.

Storage bins or rooms equipped with explosion vents of sufficient area to release any explosion to the outside atmosphere in accordance with NFPA and local codes.

All dust producing or dust agitation machinery, such as grinding mills and separators, and all elevators, elevator legs, spouts, hoppers, and other conveyors shall be provided with casings or enclosures maintained as nearly dust-tight as possible.

Dust Collection System Design

All machinery and equipment by which any explosive or flammable dust, waste, grindings, wood, filings, lint, litter, or trimmings are produced shall be permanently grounded, and shall be equipped with an efficient, approved dust collection system which shall maintain the premises free from floating lint or dust.

Belt type conveyors shall be equipped with hooded air sections, approved static collectors, and shall be adequately grounded.

Dust collection systems from grinding and other machines which may produce sparks shall not be combined with collecting systems handling linty or other readily flammable dusts.

The use of a trap at the junction of a hood or a branch duct may be permitted by the Chief provided it is not permitted to fill up with dust.

Approved magnetic separators of the permanent magnet or electromagnetic types should be installed at those points where combustible materials which may contain ferrous particles enter the system. The separators shall be of sufficient size to insure the removal of all ferrous materials passing by them.

Fan Location

Systems collection highly flammable or explosive dust shall, wherever practicable, be so arranged that the fan is on the clean air side of the collector. Systems shall be operated with approved equipment for removing stock from the air stream before it reaches the fan. Readily ignitable materials shall not pass through the fan unless the fan is constructed and installed in a manner approved therefor.

Housings of fans a blowers shall be constructed so as to prevent contact with blades thereof or accumulation therein of grease from bearings. All fan blades shall be of nonferrous or nonsparking materials. Where there is a possibility of solid foreign material passing through the fan that would produce a spark, both the rotating element and the casing shall be nonsparking material.

Ducts Design

1. Metal ducts shall be constructed of sheet metal in accordance with Table 53-A.
2. Buried ducts may be concrete or ceramic materials, cast iron, or cement-asbestos and shall be properly protected to prevent mechanical injury.
3. All sheet metal elbows and bends shall be made from material at least two gauges heavier than is required for straight duct work of the same diameter except that for No. 14 gauge and heavier, the elbows and straight duct work may be of the same gauge.
4. Round sheet metal elbows shall be of at least five-piece construction for ducts six inches in diameter or less, and of seven-piece construction for larger ducts, with a throat radius

equal to one and one-half times to two times the duct diameter unless space prohibits the use of such long bends. In place of long radius elbows specified above, rectangular elbows, venture-shaped elbows or other bends of equivalent low resistance design may be used.

5. The main suction duct should receive only one branch in a section of uniform area, whenever space permits, and in no case shall it receive more than two branches in such a section.
6. The inlet area of the fan or exhaust duct shall be equal to or greater than the sum of the areas of all the branch ducts.
7. Every branch duct shall connect with the main duct at top or side at an angle not exceeding 45 degrees, inclined in the direction of the airflow.
8. The main suction and discharge ducts shall be made as short as practicable. To provide access for sweeping and cleaning, ducts should not be less than six inches above the floor at every point, and not closer than six inches to any ceiling.
9. Every duct shall be kept open and unobstructed throughout its length, and no screen shall be placed in it. The use of a trap may be permitted as provided by codes.
10. Main ducts of systems handling materials which form an explosive mixture with air shall be run on the outside of the building where practicable, with ducts from each operation and each floor passing out directly through the wall and discharging into the main ducts.
11. Additional branch ducts shall not be added to an existing system without redesigning the system. Branch ducts shall not be disconnected nor unused portions of a system be blanked off without providing orifice plates to maintain required airflow.

Separating and Collecting Equipment

1. Cyclones, condensers, cloth screen and stocking arrestors, centrifugal collectors, and other devices used for the purpose of separating solid materials from the air stream in which it is carried, and hoppers, bins, silos, and vaults for collecting the solid materials so separated, shall be provided and maintained as hereinafter specified.
2. Separating and collecting equipment shall be designed and constructed to withstand anticipated explosion pressures, due consideration being given the reduction in pressure afforded by adequate explosion relief vents.
3. Every dust collecting system, blower, and conveyor shall discharge into an approved vault equipped with an automatic damper or choke in the discharge pipe, or into a noncombustible container outside of the building.
   In small shops, dust collecting systems may discharge into metal vaults which are vented to the outside and are located within the building. Where such collectors are of insufficient strength to resist maximum-calculated explosion pressure, they shall be located close to exterior walls to facilitate explosion relief venting.
4. Cyclone dust collectors shall vent outside of the building or to a fire-resistive, explosion vented, dust- tight room conforming to requirements of the Building Department. Supports shall be of steel, masonry, or concrete. Clean-out doors shall be provided.

5. Separating or collecting equipment shall be located at a safe distance from combustible construction or unprotected openings into buildings, and in any case not closer than a distance equal to its width or diameter.
6. Discharge ducts shall not come within 18 inches of combustible material and shall terminate at least two feet above any roof of combustible construction, and two feet above any openings within 10 feet.
7. Delivery ducts from cyclone collectors shall not convey refuse directly into the fireboxes of boilers, furnaces, (including Dutch Ovens), refuse burners, incinerators, etc.
8. Delivery of stock from separator, cyclones, or other collection equipment to storage receptacles shall, wherever practicable, be accomplished by means of gravity through tightly fitted ducts.
9. Where refuse is to be used as fuel, the discharge system from the storage receptacle or intermediate feed bin shall be so designed that either by means of a choke feeder or choke conveyor, a positive cut-off is provided to prevent a flashback from the furnace. The installation of a steam spray in the duct to the furnace, blowing steam in the direction of the fuel flow, shall be installed where practicable to provide an added safety factor in preventing a flashback.

Explosion Relief Venting

1. Explosion relief vents shall be provided on all duct systems sued for conveying materials which form an explosive mixture with air. Explosion vents shall have a cross sectional area not less than the cross sectional area of the duct vented. Vents should be provided where direction of flow is changed and at the maximum elevation in the system.
2. Explosion relief vents for ducts shall lead by the most direct practical route to the outside of the building and shall not deviate more than 22½ degrees from the direction of the duct from which they lead. Such vents shall not pass through adjoining buildings unless designed to resist maximum explosion pressure.
3. Storage bins or other receptacles which contain materials which form an explosive mixture with air shall be provided with adequate explosion relief vents.
4. Explosion relief vent openings shall be provided with diaphragms fitted with cutters to accelerate rupture or equivalent means of relieving pressure. Where necessary, the discharge opening of the relief vent may be protected against the weather in an approved manner.
5. Explosion relief vent openings shall not be connected to chimneys or duct systems used for other purposes.

Electric Lighting and Equipment

All electrical equipment within any such storage bin, vault, or storage room shall be installed and maintained in accordance with NFPA and local codes for a dust or lint hazardous location, depending upon the material handled.

Fire Protection

Buildings, rooms, conveyor systems, bins, dust collectors, and processing equipment shall be provided with fire protection equipment in accordance with NFPA pamphlets and authorities having jurisdiction.

47 Atmospheric Tanks

In this chapter we will speak of codes and regulations regulating all atmospheric tanks which are incorporated as an integral part of a transfer or process system used for handling or storing hazardous substances. Such provisions provide necessary safeguards to carry out the purpose and intent of this chapter. Whenever the term "tank" is used in this chapter, it shall mean "atmospheric tank."

Aboveground Tank. Any atmospheric tank installed above the surface of the ground and used for the storage, handling, or processing of a hazardous substance.

Hazardous Substance. All of the following liquid, solid, and gaseous substances, unless the State Department of Health Services, in consultation with the State Water Resources Control Board, determines the substance could not adversely affect the quality of the waters of the State:

1. Substances on the list prepared by the Director of the Department of Industrial Relations pursuant to Section 6382 of the Labor Code;
2. Hazardous substances, as defined in Section 25316 of the Health and Safety Code;
3. Any substance or material which is classified as a Class I, II, or IIIA liquid.

Underground Tank. Any atmospheric tank which is partially or entirely below the surface of the ground and is used to store or process a hazardous substance including any connected piping and appurtenances.

1. No person shall erect, install, relocate, add to, alter, or abandon any tank unless a valid Permit as required by Division 5 of this Article has been issued.
2. No person shall use or operate any atmospheric tank unless a valid Permit as required by Division 4 of this Article has been issued.

Every owner or operator having charge of any atmospheric tank which has an opening exceeding six inches in any lateral dimension, or which has a depth in excess of three feet, shall secure the tank from unauthorized entry. Provisions shall be made to secure unauthorized discharge from any atmospheric tank by locking valves in a closed position or by providing and maintaining equivalent protection as approved by the Chief.

1. Tanks requiring application of heat may be heated by hot water, low pressure steam coils within the tank, electrical heating units installed on the shell of the tank, or other method approved by the Chief.
2. When tanks require the heating of flammable or combustible liquids, the provisions of local fire and building codes shall apply.
aboveground tanks shall be constructed in accordance with the provisions of this Section.
3. Shop-built aboveground tanks shall be constructed in accordance with the provisions of local codes and standard.
4. Field-erected aboveground tanks shall be constructed in accordance with the provisions of local codes and standards.

The foundation and supports for aboveground tanks shall comply with the provisions of this Section. This Section is not intended to prohibit the use of concrete, sand, or gravel pads of standard construction upon which large storage tanks are normally placed.

1. Tanks which are located outside of buildings shall rest directly on the ground or on foundations or supports of non-combustible materials. Tanks shall meet seismic design requirements and compaction of the soil for footings and/or foundations in accordance with national and local codes.
2. Structural members providing support for any tank having a capacity exceeding 600 gallons of hazardous materials outside of any building shall have at least one-hour fire resistive protection or be protected with an automatic sprinkler system constructed in accordance with national and local codes.
3. Structural members providing support for any tank which is within a building and has a capacity of more than 120 gallons of hazardous materials shall be provided with structural supports which have at least two-hour fire resistive protection in accordance with national and local codes, or be protected with an automatic sprinkler system.

Location of Tanks

1. Vertical cone roof tanks constructed with a weak roof-to- shell seam shall be separated from buildings and property lines which are or can be built upon, including the opposite side of a public way, by a distance equal to the tank diameter, except that such distance shall not be less than 10 feet. Distance from streets, alleys, or public ways shall be equal to one-third of the tank diameter, except that such distance shall not be less than five feet.
2. Floating roof tanks and tanks for the storage of Class I, II, or IIIA liquids shall be separated from buildings and property lines which are or can be built upon, including the opposite side of a public way, by a distance equal to one-half the tank diameter, except that such distance shall not be less than five feet. Distance from streets, alleys, or public ways shall be equal to one-sixth of the tank diameters, except that such distance shall not be less than five feet.
3. Except as provided in Subsections A or B of this Section, every aboveground tank shall be located in accordance with local codes.

4. The distances set forth in Subsections A and C of this Section may be reduced 50 percent but not less than five feet whenever all of the tanks within any one diked area or drainage area are used exclusively for the storage of Class IIIB liquids.
5. The distances set forth in all Subsections of this Section shall be doubled for tanks storing flammable or combustible liquids having boil over characteristics.
6. The Chief may require additional spacing for tanks storing Class I, II, or IIIA liquids stored at temperatures exceeding 212°F.
7. Every aboveground tank for the storage of unstable liquids shall be located as required by the Chief, but such distances shall not be less than 25 feet from the property line of any property which has a building thereon, from the nearest side of any public way or from the nearest side of any building.
8. When tanks are installed in three or more rows, in irregular patterns, or of unusual heights in proportion to the diameter of the tank, greater spacing or other protection shall be provided so that all tanks are accessible for firefighting purposes.

Spacing Between aboveground Tanks

1. When the diameter of one tank is less than one-half the diameter of any other tank, the shell-to-shell distance between the two tanks shall not be less than one-half the diameter of the smaller tank, except that such distance shall not be less than three feet.
2. The shell-to-shell distance between any two tanks shall not be less than one-sixth of the sum of the diameters of the two tanks, except that such distance shall not be less than three feet.
3. For unstable liquids, the distance between any two tanks shall not be less than one-half the sum of their diameters.

Control of Spills, Leaks and Overflows

The accidental discharge of hazardous substances from aboveground tanks shall be diverted by means of a drainage system or shall be contained by dikes. These provisions may be altered at the discretion of the Chief when the tanks under consideration do not constitute a hazard.

Where protection is not provided by natural topography and where control of accidental discharge is by means of a drainage system to a remote impounding basin, such system shall comply with the provisions of this Section.

1. Each tank shall be installed in such a manner so as to provide a consistent slope of not less than one percent away from the tank toward the impounding basin.
2. The impounding area shall have a capacity not less than that of the largest tank that can drain into it.
3. The drainage system shall not discharge to adjoining property, public waterways, public sewers, or into adjoining common diked areas.
4. The drainage system shall terminate in a detached impounding basin or other approved containment. This containment area and the route of the drainage system shall be so located that if unauthorized release occurs, the release will not endanger adjoining property or other tanks.

5. Any group of tanks containing non-compatible hazardous materials shall have drainage systems and impounding basins designed to avoid potential intermixing.
6. The drainage system or channels shall not interfere with Fire Department Access. The drainage system may be required to be covered by an approved noncombustible grate or other approved protection.
7. Drainage system control valves shall normally be in the closed position and located outside the diked area.
8. Floating Roof Tanks: Roof drain valves shall be identified "ROOF RAINWATER DRAIN" on a sign with letters not less than three inches high on a contrasting background.

Where control of accidental discharge is by means of a diked area, such diked area shall comply with the provisions of this Section.

1. The net volumetric capacity available to a tank or group of tanks within a common diked area shall be not less than 100 percent of the largest tank enclosed by the diked area.
2. The volume of the largest tank up to the required height of the dike wall may be considered as part of the available capacity of the diked area.
3. No part of the volume of tanks other than the largest tank shall be considered as part of the available capacity of the diked area.
4. Where a separate impounding basin is used to reduce the required capacity of a diked area, drainage sufficient to prevent overflow of the dike and effective control of flow shall be provided.
5. The volumetric capacity of a single impounding basin may be applied to reduce the required capacity of each of the diked areas draining to it.
6. Walls of the diked area shall be of earth, steel, concrete, or solid masonry designed to be liquid-tight and to withstand a full hydrostatic head. Earthen walls three feet or more in height shall have a flat section at the top not less than two feet wide. An earth dike wall shall have minimum slope of one and one-half feet horizontal to one foot vertical and shall be well compacted and coated with concrete, asphalt, or other material approved by the Chief.
7. The walls of the diked area shall be restricted to an average height of six feet above interior grade. The minimum distance between the inside toe of any dike wall and the shell of any tank shall be not less than five feet. Concrete footing for steel, concrete, or solid masonry walls may project into this area, provided the top of the footing does not project above interior grade.
8. Any group of tanks containing non-compatible materials shall have separate diked areas designed to avoid potential intermixing.
9. Each diked area containing two or more tanks shall be subdivided, preferably by drainage channels or at least by intermediate curbs, in order to prevent spills from endangering adjacent tanks within the diked area as follows:
   a. When storing normally stable liquids in vertical cone roof tanks constructed with weak roof-to-shell seams or approved floating roof tanks, or when storing crude

petroleum in producing areas in any type of tank, one subdivision shall be provided for each group of tanks (no tank exceeding 10,000 barrels capacity) having an aggregate capacity not exceeding 15,000 barrels.

b. When storing normally stable liquids in tanks not covered in Subsection 1 of this Section, one subdivision shall be provided for each tank in excess of 2,500 barrels, and one subdivision shall be provided for each group of tanks (no tank exceeding 2,500 barrels capacity) having an aggregate capacity not exceeding 3,750 barrels.

c. Whenever two or more tanks, any one of which is over 150 feet in diameter, are located in a common diked area, intermediate dikes shall be provided between adjacent tanks to hold at least 10 percent of the capacity of the tank so enclosed, not including the volume displaced by the tank.

d. The intermediate curbs or drainage channels shall be located between tanks so as to take full advantage of the available space with due regard for the individual tank capacities. Where intermediate curbs are used, they shall be not less than 18 inches in height, and construction requirements shall be the same as for dike walls.

10. Within each diked area, drainage shall be provided at a consistent slope of not less than one percent but not more than five percent, away from tanks toward a sump, drainage system, or other safe means of disposal located at the greatest practical distance from the tank. Such drains shall normally be controlled in a manner so as to prevent hazardous substances from entering natural water courses, public sewers, or public drains. Control of drainage shall be accessible under fire conditions from outside the diked area.

11. No dry vegetation or combustible material not a part of the permanent installation nor any empty or full drum or barrel shall be permitted within a dike area.

Venting of aboveground Tanks

1. Atmospheric storage tanks shall be adequately vented to prevent the development of vacuum or pressure sufficient to distort any portion of the tank or to exceed the design pressure, which might result during filling or emptying, or changes in temperature or atmospheric pressure. Tanks storing Class I liquids shall be equipped with either venting devices which shall be normally closed when not under pressure or vacuum or with an approved flame arrester.

2. Every aboveground storage tank shall have an emergency vent which shall consist of some form of construction or device that will relieve excessive internal pressure caused by exposure fires that might cause the rupture of the tank shall or bottom. In a vertical tank, this construction may take the form of a weak roof-to-shell seam. Where entire dependence for such additional relief is placed upon some device other than a weak roof-to-shell seam, the total venting capacity of both normal and emergency vents shall be sufficient to prevent rupture of the shell or bottom of the tank, if vertical, or of the shell or heads, if horizontal. Such device may be a self-closing manway cover or one using long bolts that permit the cover to lift under internal pressure, or an additional or larger relief valve or valves.

3. The size of vents shall comply with the requirements of local codes or shall be large as the filling or withdrawal connection, whichever is larger, but in no case less than one and one-quarter inches nominal inside diameter.
4. Tank vents shall be readily accessible for inspection and shall be properly maintained at all times.

Grounding and Bonding

All aboveground tanks not inherently grounded shall be electrically grounded or bonded as approved by the Chief.

Stairs, Walkways, and Platforms

Stairs, walkways, catwalks, and similar equipment on or adjacent to aboveground tanks or within dike walls shall be of noncombustible construction and in locations approved by the Chief.

Gauging devices for aboveground tanks shall be of a type and design that will not continuously expose the vapor space above the liquid level in the tank, and will minimise liquid loss should the device be damaged mechanically or by fire.

Sources of ignition shall not be within 25 feet of any aboveground tank containing:

1. Class I liquids.
2. Class II or III liquids at temperature above their flashpoints unless safeguarded as approved by the Chief.

Tanks inside Buildings

In addition to the general requirements set forth in this division, aboveground tanks located inside buildings shall conform to the following:

1. Tanks located inside buildings shall be limited to 500 gallons individual capacity unless additional capacity is approved by the Chief.
2. No tanks shall be located, installed, maintained, or used inside any building except in an approved location and in conformance with the following specifications:
    a. No tank shall be located within five feet of combustible materials nor within five feet of any exit.
    b. Tanks with not more than 500 gallons of Class II or III liquids may be installed below grade in basements of buildings if located in a fire-resistive room provided with automatic fire protection and an approved detection system. Such room shall be capable of containing 100 percent of the total capacity of all tanks. Tanks shall be installed in compliance with the specifications for aboveground tanks contained in local standard.
    c. Vapors from normal and emergency vents of tanks within buildings shall be exhausted at a safe location outside the building.
    d. The use of weak roof seams on tanks within buildings is prohibited.

3. Every tank exceeding 10 square feet of exposed surface area shall be equipped with a tight-fitting cover or an approved automatic closing cover, or an approved local ventilation system. All covers shall be kept tightly in place at all times other than during periods of filling, withdrawal of contents, or cleaning.
4. All open top tanks containing flammable or combustible liquids and exceeding 150 gallons capacity or 10 square feet surface area shall be provided with approved drains curbs, overflow pipes, and similar protection in accordance with the provisions in Division 81 of this Article.
5. Connections to horizontal tanks shall be through the top. Fill, return, and similar pipes shall be extended below the level of the suction pipe intake or shall be trapped inside the tank in a manner to prohibit the escape of vapors.
6. Connections to vertical tanks shall be through the top wherever practicable, and otherwise through the side.
7. Atmospheric tanks containing flammable or combustible liquids shall also be subject to the following conditions:
   a. The inlet of the fill pipe shall be located outside of the building and not less than five feet from any building openings. The fill pipe shall terminate within six inches of the bottom of the tank.
   b. Tanks located in buildings, under buildings, or where the tank location is not visible from the fill opening, shall be equipped with closing devices installed in the fill line in a manner to prevent filling the tank to more than 95 percent capacity. Tanks equipped with an overflow pipe at least one pipe size larger than the fill pipe discharging to an approved location.
   c. Withdrawal connections through the side of a tank below the liquid level shall be equipped with an automatic closing heat-actuated device. Engine fuel tanks installed in conformance with codes are exempt from the requirements of this Subsection.

Ventilation in Buildings

1. Every room or building in which tanks are located or in which Class I, II, or IIIA liquids are manufactured, used, processed, or stored shall be equipped with natural or approved mechanical means of ventilation sufficient to prevent the formation or retention of flammable vapors.
2. Explosion venting methods and devices approved by the Building Department shall be installed in every room in which flammable vapors present a potential explosion hazard.

Reconditioning of Used Tanks

Tanks which are to be reconditioned or abandoned by welding, cutting, or other "hot work" methods shall be rendered gas-free before any welding or cutting operations are conducted. All operations in connection with the reconditioning of used tanks using such methods shall be carried on under supervision of experienced personnel and in compliance with local codes and standards.

Atmospheric tanks containing hazardous materials may be insulated by the application of insulating material approved by the Chief. Insulating materials shall be fire resistant and shall not decompose rapidly if exposed to fire.

Engine Fuel Tank

1. Every engine fuel tank shall be approved by the chief and equipped with an approved emergency vent of sufficient size to properly relieve any internal pressure which might be created from exposure to heat or fire.
2. Every engine fuel tank shall be so installed that the top is lower than the fuel pump or carburetor or any such engine.
3. No fuel supply pipe shall be installed except in the top of the tank. The fuel supply system shall be installed so that broken pipes or loose connections will not permit leakage from the fuel tank.
4. Unless located in an approved fire resistive room, any fuel tank located inside of, or on the roof of a building shall be limited to 25 gallons capacity for gasoline and 60 gallons capacity for diesel.
5. When installed within two feet of an exhaust manifold or pipe, the fuel tank shall be protected against heat by a metal baffle or equivalent insulation.
6. Every pump which supplies any engine fuel tank shall be provided with an interlock device so that the pump may operate only when the engine being supplied is in operation.

Warning Placards and Signs

Every aboveground tank used for the storage of hazardous substance shall have conspicuously posted warning placards and signs in accordance with the provisions of this Section. Warning placards shall be designed and constructed as specified in locals standard and shall be posted on at least two side of any such aboveground tank. Any tank containing Class I liquids at any temperature, of Class II or III liquids at temperatures above their flash points shall also have conspicuously posted thereon a sign in letters not less than three inches high on contrasting background reading "NO SMOKING OR OPEN FLAME WITHIN 25 FEET."

In lieu of posting each tank in areas containing more than one tank, the area itself may be posted in a manner approved by the local codes.

Fire Protection for Tanks

1. Every aboveground tank having a liquid surface area of 10 square feet or more shall be provided with approved fire protection equipment, an inserting system, or otherwise safeguarded against fire or explosion in a manner approved by the Chief.
2. Portable fire extinguishers shall be of a size, number, and located as required in Division 140 of this Article.

Underground Tanks

In addition to applicable requirements of this Division, the following sections shall regulate underground atmospheric tanks and appurtenances used for the storage or processing of hazardous substances.

The provisions of these Sections shall not include structures such as sumps, separators, storm drains, catch basins, oil field gathering lines, refinery pipelines, intrastate or interstate pipelines, lagoons, evaporation ponds, well cellars, separation sumps, lined or unlined pits, aboveground storage tank spill containment areas, or final interceptors in industrial waste pretreatment systems.

Definitions

The following words and phrases whenever used in these sections shall be defined as follows:

- Existing Facility: Any underground tank containing a hazardous substance which was installed or placed in service prior to January 1, 1984.
- Facility: Any underground tank or group of underground tanks containing hazardous substances which are located at a single site.
- New Facility: Any underground tank containing a hazardous substance which was installed or placed in service on or after January 1, 1984.
- Primary Tank: The portion of an underground tank which comes into immediate contact on its inner surface with the hazardous substance being contained.
- Secondary Containment: A system of containment external to, and separate from, the primary tank. One secondary containment may serve as secondary containment for more than one primary tank.
- Single Walled: The construction of tank walls consisting of one thickness of material. Laminated, coated or clad materials shall be considered as single-walled.
- Unauthorized Release: Any emission or escape of any hazardous substance which does not conform to the provisions of this Division, unless such release is in accordance with the release regulations of the South Coast Air Quality Management District and California Air Resources Board, with a National Pollutant Discharge Elimination System (NPDES) Permit, or with waste discharge requirements established by a Regional Water Quality Control Board.

New Underground Tank Requirements

Every new underground tank shall be provided with and meet the following minimum requirements.

1. Secondary containment in accordance with local codes.
2. Construction, corrosion protection, and testing of underground tanks in accordance with local codes.

Secondary Containment

1. Every new facility shall be designated and constructed with secondary containment.
2. Secondary containment shall be compatible with any hazardous substance contained within a primary tank or within several primary tanks served by a single secondary containment. Hazardous substances shall not be intermixed so as to cause a fire, explosion, or the production of toxic or poisonous gas. Hazardous substances which are not

compatible with each other shall have separate primary tank and secondary containment systems.

3. Secondary containment shall be large enough to contain at least 100 percent of the volume of the primary tank.
4. Secondary containment which includes multiple primary tanks shall be capable of containing at least 150 percent of the volume of the largest primary tank.
5. When there is a possibility of water intruding into the space between the primary tank and secondary containment by precipitation, infiltration, or other means, the facility shall be designed and constructed with the capability to monitor and remove such water intrusion.

Construction, Corrosion Protection, and Testing of Underground Tanks

The construction, corrosion protection, and testing of underground tanks shall comply with the requirements of local building and fire codes.

Requirements for Existing Underground Tanks

Every existing underground tank shall meet the following minimum requirements by January 1, 1985.

1. Facility inspection evaluation in accordance with local codes.
2. Integrity testing in accordance with local codes.
3. Soils investigation in accordance with local codes.
4. Monitoring systems in accordance with local codes.
5. Facility inspection and records in accordance with local codes.

Facility Inspection for Continued Use

The continued use of, and permit approval for an underground tank or facility is subject to modification or termination by the Chief whenever there has been any unauthorized release, or when necessary to comply with the minimum requirements of this Article.

In determining whether continued storage in such underground tank is suitable, the Chief shall consider the following criteria:

1. The age of the underground tank and facility.
2. The method of containment.
3. The method of monitoring.
4. The feasibility of any required retrofit.
5. The amount and type of hazardous substance stored.
6. The severity of any potential unauthorized release.

Integrity Testing of Existing Underground Tanks

Existing underground tanks and piping shall be tested for tightness by an approved method capable of detecting a minimum leak rate of 0.05 gallons per hour or less. A soils investigation analysis may be required to determine precontamination or the adequacy of past hazardous substance containment and the appropriate monitoring systems to be installed. Whenever a leak is detected, the Chief shall be notified in accordance with local codes.

Soils Investigation

To determine whether or not an unauthorised release has occurred at a facility, the Chief may require a soils analysis by soil boring, soil samples, or other approved method. The soils analysis along with documentation of the sequence of custody of the soil analysis shall be submitted to the Chief in writing and certified by a California Registered Engineering Geologist or a California Certified Geologist.

Monitoring Systems

Every underground tank shall be provided with a monitoring system designed and constructed to be capable of detecting the release of any hazardous substance.

Existing underground tanks which are in a concrete vault and the exterior surface of the tanks are accessible to visual inspection. Visual inspections shall be conducted at intervals specified by the Chief.

1. When a secondary containment system is used, the monitoring device shall be capable of detecting leakage into the space between the primary tank and secondary containment.
2. Secondary containment systems, other than double-walled tanks, which may be subject to the intrusion of water or other liquids shall be provided with a means of detection and removal.
3. All methods of monitoring, detection, and leak interception shall be certified by a California Registered Engineer and be approved by the Chief.
4. Monitoring systems shall be connected to alarms which shall be both visual and audible.
5. A monitoring system annunciator panel, when required, shall be installed in locations where responsible personnel are normally in attendance during operation of the facility. These panels shall not be installed in closets or storage rooms, nor in other isolated areas.
6. Written operating instructions and test procedures shall be located on the monitoring system annunciator panel giving a step-by-step explanation of the procedures to be used when an unauthorized release occurs or when testing the monitoring systems.

Facility Inspection and Records

The owner, operator, or permittee of a facility shall conduct or cause to be conducted monthly inspections and tests of the facility, including the monitoring systems of that facility. Records of these inspections and tests shall be maintained for at least three years and shall be available for inspection by the Chief.

Location of Underground Tanks

1. Tanks which are located in areas where they may become buoyant due to a rise in the water table, or in areas which are subject to flooding, shall be anchored in a manner to prevent their floating or unsafe movement.

2. Tanks shall be so located with respect to existing building foundations and supports that the loads carried by these foundations and supports cannot be transmitted to the tank. The distance from any part of a tank to the nearest wall or any basement, pit, or cellar shall not be less than one foot and not be less than three feet from any property line that may be built upon.
3. Underground tanks shall be separated from each other by at least one foot.

Installation of Underground Tanks

1. Excavations shall be made with proper precautions for existing foundations and/or load bearing supports.
2. Tanks shall be set on a firm foundation of at least 12 inches of noncorrosive inert backfill material such as clean sand or pea gravel well tamped in place on the tops, sides, and ends of tanks.
3. Tanks located in protected areas where they are not, and will never be, subjected to vehicular or other superimposed loads may be installed with a cover of not less than two feet of earth when approved by the Chief.
4. Tanks located where subjected to traffic shall be installed at least three feet below the lowest grade or shall be covered with at least 18 inches of noncorrosive inert backfill material such as clean sand or pea gravel well tamped in place plus six inches of reinforced concrete slab that extends one foot beyond the extremity of the tank and is designed to sustain the vehicular and other superimposed loads in accordance with local building and fire codes.
5. The fill box or opening shall be safeguarded against spills or unauthorized release where asphaltic or concrete paving is not provided.
6. Underground tanks which have a remote fill shall be equipped with automatic closing devices installed in the fill line to prevent filling the tank more than 95 percent. In addition to such closing device, there shall be an audible alarm set at 90 percent capacity. The remote fill and alarm shall be labeled and located in a manner approved by the Chief.

Venting of Underground Tanks

1. Every underground tank having only one fill opening shall be equipped with a vent pipe of not less than one and one-fourth inch nominal inside diameter. Tanks requiring vapor recovery systems shall have no less than one vent pipe with a nominal inside diameter of not less than two inches. Vent pipes shall have the capacity to prevent back-pressure development in tanks from exceeding 2.5 PSIG.
2. All Class I liquid vent pipes shall be 18 inches above the highest point of any roof of any building within 15 feet of the point of termination. Vent pipes shall terminate 12 feet above the adjacent ground level and not less than 10 feet from building openings. They shall discharge only upward or horizontally to disperse vapors away from exposures and hazardous locations.

3. Every vent pipe shall run without sag or trap from the top of the tank to which it is connected and shall be so located and directed that flammable vapors will not accumulate or travel to an unsafe location, enter building openings, or be trapped under eaves.
4. Vent pipes running in a semi-horizontal direction shall be installed to provide one-fourth inch of fall for each one foot of horizontal run toward the tank.
5. Vent lines from two or more underground tanks may be manifolded into one main vent line provided the manifold is located not less than eight feet above the topmost point of every fill pipe for such tanks. The internal cross sectional area of the line above the manifold shall equal the combined cross sectional areas of all pipes so manifolded. Tanks containing Class I liquids shall not be manifolded with tanks containing Class II or III liquids, nor shall tanks which contain non-compatible hazardous substances be manifolded.
6. Vent pipes shall be located and supported in an approved manner.
7. Vent pipes shall be unobstructed and only approved pressure relief valves or approved bail check valves shall be installed inside the tank.

Installation Inspection

Underground tanks, secondary containment, monitoring systems, and piping shall not be covered until the installation has been inspected and approved by the Chief.

Suction Delivery System Exemption

Suction delivery piping systems may be excluded from the requirement for secondary containment when the following are provided:

1. Vent piping, product return piping, and vapor recovery piping are equipped with an approved ball valve back flow protection installed inside the tank which prevents the contents of the tank from entering the piping under pressure.
2. Vent piping, product piping, product return piping, and vapor recovery piping systems run directly to tank without sag or trap and with one-fourth inch fall for each one foot of horizontal run.
3. The product suction pump is in an approved location.

Pressure Delivery System Exemption

Pressure delivery piping systems may be excluded from the requirement for secondary containment when the following are provided:

1. An approved leak detection system shall be provided for the product piping which shall automatically shut-off the pumps when a leak is detected.
2. Vent piping, product return piping, and vapor recovery piping are equipped with approved ball valve back flow protection which prevents the contents of the tank from entering the piping under pressure.
3. Vent piping, product return piping, and vapor recovery piping systems run directly to tank without sag or trap and with one-fourth inch fall for each one foot of horizontal run.

Unauthorized Release

1. No person shall cause or allow the storage of hazardous substances in a manner which causes an unauthorized release of hazardous substances or poses a significant risk of such unauthorized release.
2. Reporting — Any person who has knowledge of an unauthorized release of a hazardous substance shall notify the Chief immediately. A written report to the Chief shall be submitted by the owner, operator, or permittee within 10 days after the escape of a hazardous substance from a primary tank when such unauthorized release:
   a. Has the potential to enter groundwater.
   b. Could endanger life or property.
   c. Has been contained by secondary containment.

The written report shall provide information to the Chief relating to the ability of the permittee to contain and dispose of the hazardous substance, the estimated time it will take to achieve this, and the degree of hazard created.

The Chief may verify that the hazardous substance is being contained and appropriately disposed of, and if at any time it is determined that the permittee is not adequately containing and disposing of such hazardous substance, shall have the power and authority to initiate and direct an emergency response in order to protect the public health and/or safety. The permittee shall be liable for reimbursement to the Department for all emergency response costs incurred.

3. Cleanup — The owner, operator, or permittee shall be responsible to take all necessary steps to ensure the containment and cleanup of any unauthorized release has entered or has the potential to enter groundwater, the Chief shall immediately notify the California Regional Water Quality Control Board so that they shall assume lead agency status for the investigation and abatement of the release.

Ground Level Service Openings

All service openings at ground level over or leading to the primary tank or secondary containment shall be so designed as to be liquid tight and not allow any drainage or contamination to enter through such opening and shall be installed a minimum of one inch above adjacent ground level.

Fill Box

Every fill box shall have a product identification label attached. The fill box fixture shall have a capacity of not more than five gallons and all metallic components shall be appropriately grounded. The fill box or cap shall be capable of being locked.

Security of Access

Access to a facility shall be secured by fences and/or locks or other equivalent protection approved by the Chief.

Temporarily Out of Service

Any underground storage tank which is temporarily out of service shall be monitored and inspected in the same manner as if it were in service.

Abandonment of Underground Tanks

All abandonment work required by this Section shall be inspected by the Chief.

1. Whenever an underground tank is out of service for a period of 90 days or when there is not a valid Permit to operate, the tank shall be abandoned within 30 days after notice has been given by the Chief.
2. The abandonment of underground tanks, piping and appurtenances shall be accomplished by removing them from the ground. An alternate method may be approved by the Chief when removal from the ground would cause an unsafe condition or would be impractical.
3. The owner, operator, or permittee of any underground tank which is to be abandoned, or the agent designated by such person to perform the abandonment work, shall notify the Chief of the time and place of such abandonment no later than 4:00 p.m., two working days or 48 hours prior to the proposed date of abandonment operations.
4. Before any underground tank is transported, it shall be inserted in such a manner as to render it free of flammable vapors.
5. The permittee shall establish to the satisfaction of the Chief that no unauthorized release of hazardous substance has occurred.
6. Secondary containment, monitoring, and detection systems shall be abandoned in a method approved by the Chief.

REPAIR OF UNDERGROUND TANKS

Underground tanks may be repaired one time in a method approved by the Chief and in accordance with the following provisions:

1. Before such repair, a soils investigation shall be conducted to determine if an unauthorized release has occurred.
2. A permit shall be issued in accordance with Division 5 of this Article.
3. After such repair, the tank shall be tested in a method approved by the Chief.

Reinstallation

No person shall reinstall any used or refurbished underground tank unless the same shall have been inspected and precision tested in accordance with local codes.

50 Fire Protection for Chimneys

Most people think of their chimneys as nothing more than the exit route for harmless smoke from their fireplaces or wood stoves. They assume that if the chimney isn't crumbling or visibly damaged on the outside, it's in good shape. But the danger of chimney fires comes not only from structural problems, but more often from poor maintenance.

Though wood stoves and fireplaces are the cause of most chimney fires, all fuel-burning appliances require sound, clean well-maintained flues. Even gas appliances put significant moisture up the flue that can cause serious problems, including rapid deterioration of masonry chimneys.

While cholesterol is the culprit that builds up in your arteries to threaten heart attack, fuel oil and coal fires build up soot on chimney walls. In the case of fireplaces and wood stoves, creosote, a natural by product of burning wood, is the offender. And though it's very flammable and therefore dangerous, it's also removable, and to some extent manageable. The type of wood you burn, the way in which you burn it, and the size of your chimney are all factors you can control to reduce creosote buildup and help avoid the hazards of chimney fires.

Trees are either deciduous or evergreen, generally referred to, respectively, as hardwoods and softwoods. The most preferable for burning are the "hard" hardwoods: maple, birch, and oak. The dense wood produces concentrated energy from the fires for a longer period of time, resulting in the least amount of creosote build-up. On the other hand, softer hardwoods, such as cottonwood and aspen, are less dense and will burn faster.

Evergreens, like hemlock, spruce, and pine, produce beautiful fires because they burn quickly and brightly, but the resins in them produce gases and by products that cause excessive creosote build-up when burned slowly. If you must use these softwoods, build small, hot fires. Be sure you use dry, well-seasoned wood.

Now is the time to think about next winter's wood supply. Season wood at least six, preferably 12 to 18, months before use. Green wood gives off moisture, gases, and resins as it heats and causes creosote to build up quickly in your flue. Green wood will sizzle and boil as it tries to burn; it's difficult to ignite and can cause the fire to die if air supply is reduced. It is heavier and looks different from well-seasoned wood that often shows cracks radiating outward toward the bark, like wheel spokes. Green wood won't show this pattern of cracks. It should also cost about 25 percent less than dry wood.

Safe Burning

If you have a wood stove, be sure to read the operating instructions about your model's optimum performance routine. If you're keeping the wood stove burning all day to maintain the heat level in your home, open the damper, stoke the fire, and burn a hot fire (500 degrees) for an hour twice daily.

This procedure helps avoid creosote buildup by burning off the small amount of accumulation daily. It also serves to remove the water in the deposit, making maintenance an easier task. You should also keep your fire burning above 225 degrees. Below that temperature it's a "cold" fire that causes the resins to condense and produce creosote. You'll always know the temperature of your fire if you use a stovepipe thermometer. Inexpensive ones that attach magnetically to your stovepipe are available from your hardware store. Wood stove specialty stores may offer a more accurate probe-type thermometer.

Spring and fall are the times when creosote build-up occurs most frequently. When the weather is cool but not cold, a minimal fire is needed for heat. There is a tendency to let the fire smolder for long periods. Smoldering means incomplete combustion, and therefore, creosote build-up.

When you don't need much heat, try to build quick, hot fires, or use your hone's central heating system for warmth.

The Annual Inspection

In addition to proper burning techniques, you should perform an annual inspection. A professional inspection by a chimney sweep or mason will cover both structural and maintenance issues. You should consider working with a chimney sweep certified by the Chimney Safety Institute of America (CSIA).

Such an inspection will ensure, for example, that the flue size is in proper proportion to the appliance. A large fireplace chimney that now serves a wood stove with a smaller flue requirement is especially vulnerable to creosote buildup. To prevent this from happening, liners are available that are fabricated in stainless steel or cast on site to fit the appliance and the chimney. A chimney sweep should be called whether you have a prefabricated or masonry fireplace, free-standing wood stove, hearth stove, or insert.

Your yearly inspection should examine the cleanliness and the structural soundness of both your wood-burning appliance as well as your chimney (see safety checklist). It should also address home safety issues such as fire extinguishers, smoke detectors, and your family's fire escape plan.

With proper burning techniques and dry hardwood, you may not need to clean your chimney more often than once a year, though only a thorough inspection can tell you that. When your chimney does need cleaning, you can either call a reliable professional or do it yourself. If you choose to do it yourself, the National Chimney Sweep Guild recommends that you clean the inside of the chimney by using a good stiff chimney brush (available from your local hardware store) connected to fiberglass rods or a rope lowered from the top.

Tape an old sheet across the fireplace opening, or make sure the wood stove is in the airtight position to prevent soot from spreading throughout the room.

Safety Checklist

Be sure your annual inspection includes the following areas:

Fireplace

1. Hearth protection
2. Smoke chamber
3. Firebox/grate
4. Spark screen or doors
5. Ash container
6. Damper
7. Tools and gloves

Chimney

1. Brickwork
2. Mortar
3. Flue tiles
4. Crown/wash
5. Chimney cap/spark arrestor
6. Flashing

Wood Stove

1. Stove pipe condition
2. Stove gasketing
3. Thimble, flue connection
4. Installation/clearances

A clean chimney and controlled fire will reduce the chance of a chimney fire. However, if a chimney does catch fire, close the draft control on the stove and call the fire department immediately. Don't take a chance by letting a chimney fire burn itself out.

The Chimney Safety Institute offers two booklets, "CSIA Certified Chimney Sweeps and You" and "The Facts about Chimney Fires: Causes & Cures,". For a copy please write, specify which booklet or both:

1. Chimney Safety Institute of America 16021 Industrial Dr., Suite 8
2. Gaithersburg, MD 20877.

Safe Use of Your Wood Stove

1. Keep two inches of sand or ash on the bottom of the empty stove to reduce radiation and high temperatures underneath the stove.

2. Always open dampers before starting a fire in a wood stove. Always use paper and kindling wood. Do not use lighter fluid, kerosene, or gasoline to kindle a fire.
3. Do not place any combustible materials within the specified clearance area or store dry wood near or under the stove.
4. Be careful of burning embers from the stove when adding wood.
5. Before wood will ignite and burn, it must be heated to combustion temperature and the moisture in the wood must be driven off. During the heating process, wood gives off condensable gases and tars. So, with an improperly built fire, gases and tars will condense inside the chimney, build up as creosote, and may lead to a chimney fire. Use only seasoned, dry wood. Don't place green or wet wood on top of a low fire. This will simply cause the distillation of condensable materials that will then build up in the chimney.
6. A roaring fire can cause chimney fires or overheat the stove and cause it to crack.
7. Use a covered metal container to hold ashes you've shoveled out and allow them to cool before disposing of them.
8. Burn artificial logs in a fireplace only — never in a stove.

51 Fire Protection for Aircraft Cabin

Inflight cabin fires in aircraft are rare, but when they do occur, the results are often catastrophic. Until recently, the very infrequency of this menace has allowed the Feberal Aviation Administration (FAA) as well as its counterparts around the world to pay slight attention to the threat in order to focus on more visible safety problems.

However, several devastating accidents involving inflight fires in corporate aircraft extinguished the apathy. Those accidents coupled with several high-visibility, tragic airline cabin fires during the last decade, prompted the FAA to spend more money and to devise a more sharply focused program on aircraft interior fire prevention.

United States Congress sponsored what has become the Aviation Safety Research Act of 1988. The law required the testing of aircraft interior materials for fire resistance and established standards for flammability, smoke and toxicity (FST).

Another important contribution to the cause of reducing the risk of fire due to the flammability of aircraft interiors was the first International Conference for the Promotion of Advanced Interior Materials. Organized by the FAA's relatively new Fire Safety Branch, the meeting was held in February 1993 at the FAA Technical Center in Atlantic City, New Jersey. Attendees included several hundred fire protection specialists from a dozen countries, along with aircraft operators and representatives of user groups.

Among the speakers were officials from civil and military aircraft manufacturers, research organizations and aviation regulatory and oversight organizations from around the world. Ronald Ashford, representing the 20 European country members of the Joint Aviation Authorities (JAA), spoke of the need for greater awareness of existing fire hazards in FAR Part 23 and 25 aircraft and reported on the progress the JAA and the FAA are making in adopting common fire safety standards.

While there is still much to be done in achieving complete harmonization between the JAA and the FAA, Ashford said he anticipated the JAA Joint Aviation Regulations (JARs) would be published by early this year for implementation in 1995. Whether the FAA will adopt similar regulations on a similar timetable was not clear, but all countries and all aircraft operators will benefit from the dramatic improvements in fire-resistant materials that were reported on at the Conference.

Today's technology has produced a Most of materials suitable for aircraft interiors — including panels, overheads, carpets and seat materials that can withstand a temperature up to 2,000°F without measurable FSTs. In a field with the caveat, "Given enough heat, anything can burn," this is news indeed. With these "miracle" products under development, achieving the Conference's goal of helping the fire prevention industry develop a truly fireproof cabin by 2010 seems well within reach.

In the meanwhile, the director of Fire Safety Branch at the FAA Technical Center has recommended the following:

1. Airline and business aircraft operators should pay more preflight attention to fire prevention.
2. Flight attendants should be prepared to take immediate containment steps in the event of an inflight fire.

FAA claims that there have been repeated instances of untimely and ineffectual responses to cabin fires which had allowed the hazard to increase markedly. In fact, one of the most important responses to indications of inflight fire is to act positively and promptly to the threat and to alert passengers to the possible severity of the situation.

Consider the following examples of two corporate aircraft accidents involving in flight fire within the last few years. Each illustrates the importance of crews being familiar with emergency procedures and following them to the letter, including informing passengers of the severity of the problem. In both instances, crew members alerted ground services, declared an emergency and undertook rapid descent and landing.

Apparently, however, they failed to inform passengers about the operation of the main door or the use of emergency exits. One crash landing was survivable, but the crew was unable to immediately assist with getting the main exit door open. Passengers had no recourse but to struggle with the door release, and they couldn't get it to budge. They were ultimately rescued from the outside, but only after the passage of precious minutes.

In the second case, the main exit door of the aircraft had been damaged, and these passengers, too, wasted valuable time trying to figure out how to open it. Eventually they were able to exit through a fracture in the rear fuselage.

Prompt Response Necessary

FAA also indicated that all too often, crew members have a passive attitude toward inflight fires. They seem to feel that since such events are random, unpredictable and rare occurrences, there is not much one can do to prepare.

In fact, information provided by the National Transportation safety Board (NTSB) indicated that most inflight fires occur in either the lavatory or the galley, and both situations call for immediate and well-structured crew response.

In the lavatory, "over–servicing" (the term the fire prevention business uses to refer to careless and even excessive use of the system) can result in liquids and high voltages coming together to cause a short and arcing that can ignite surrounding materials. And most lavatories are constructed of tough synthetic materials that are likely to burn easily. The rule of thumb is that if a fire is suspected in the lavatory, pull off all electrical power to that area and use an extinguisher only in the manner specified in the instruction manuals.

In the galley, the same rules roughly apply since the ingredients for accidental fire are glaringly present. All it takes is flammable wrapping materials or containers, a little heat and haste mixed with a serving of turbulence. Spilled liquids — some corrosive and some with high alcohol content — can mean trouble in the vicinity of high heat, flame or burning tobacco.

Trash in containers that can burn and produce toxic smoke is an ever present hazard in many aircraft. Some major completion centers such as Kansas City (KC) Aviation of Dallas and Duncan Aviation of Lincoln, Nebraska offer a fire-smothering device for lavatory and galley trash containers, and it automatically activates in case of fire. Also, all major completion centers will install a master switch system that can cut all electrical power to either the galley, lavatory or the cabin.

Symptoms Spell Trouble

The NTSB also cited the temptation of some crews to ignore indications of fuel leaks — apparently in the hope that such problems will go away if they are ignored long enough. A recent edition of the National Business Aviation Association (NBAA) "Maintenance and Operations Bulletin" (MOB) reported a case that makes the point. A corporate aircraft flew for almost a year with the faint but persistent smell of fuel in the cabin area. The crew had tried repeatedly to locate a leak, but finally concluded the smell was the result of fuel residues on lines and fixtures.

Finally, during a major inspection, diligence paid off, and the source turned out to be an almost hidden 90-degree elbow at the entry point of a line connecting the right-hand tank to the fuel selector. The joint had been cross-threaded, resulting in a slow, continuous lead that left no visible residue under the aircraft. However, plenty of fuel was exposed to trigger and explosion.

"Thanks to some conscientious work, a flying bomb was defused," noted the MOB.

Hazardous Materials

Although difficult to monitor, it is essential that the crew or other aviation department personnel alert passengers prior to flight to the potential hazards of bringing what may seem to be "harmless" materials on board. Some operators make preflight announcements warning of possibly hazardous materials. Others distribute prepared handouts that remind passengers to speak to crew members about the possibility of normally inert chemicals or other materials igniting if subjected to wide temperature changes, pressurization or turbulence.

A recent case involved usually harmless chemicals that were to be a part of a sales presentation. The substances were stored in unprotected and loosely-tied packages in a baggage area. Turbulence caused spillage, and the mixing of two chemicals resulted in a fire.

In another case, seafood stored in salt water in a baggage area leaked onto wires below the floor. The frayed wiring insulation caused shorting and fire, precipitating an emergency landing and — in this instance — a successful one.

Crews should also remember that the period immediately following major maintenance is a time to be extra wary of fire hazards. In spite of careful post-maintenance cleanup, battery or hydraulic system spills might be left unwiped, or tools and other material might be left behind, resulting in insulation chafing and exposed wires.

Hidden Fires and Aging Aircraft

In fact, wiring is a major contributor to aircraft fires, and insulation chafing is the most common mode of wire failure, according to George A. Slenski of the Wright Laboratory Materials Directorate at Wright Patterson Air Force Base in Dayton, Ohio. Insulation is also susceptible to becoming brittle and frayed over time and can gradually become "carbonized." This deterioration can prompt arcing and precipitate fire.

Unfortunately, such fires can occur in almost any hidden corner of the airplane and can be difficult to locate and extinguish. The best solution is to replace older wiring at inspection times or when new equipment is being installed.

While there is no FAA requirement that dictates the replacement of aging systems unless strong evidence of an unairworthy condition exists, many completion and maintenance centers make it a practice to inspect older wiring or to replace it as necessary.

The most important advances in fire-retardant products for aircraft have been in the development of insulation. Now wiring insulations are strongly flame-resistant and are almost immune to the softening and hardening that takes place during wide temperature variation in older systems.

Furthermore, these next generation "hybrid" insulation materials are expected to cost no more than comparable insulations currently used on aircraft.

Present FAA rules do not require Part 23 or Part 25 aircraft to upgrade older interiors to current standards unless an aircraft is undergoing "substantial" refurbishment. But most modification and completion centers for business aircraft invariably remind owners of older aircraft about the benefits of upgrading at least the lavatory and galley areas to current standards. Many customers follow through on that advice because they recognize the dramatic improvements that have been made in developing fire-resistant and fireproof materials.

One of the most important advances in the development of fire-resistant materials for aircraft has been the establishment of highly sophisticated standardized testing. The tests are a far cry from the relatively primitive evaluations that were introduced in the 1960s. I n those days, most interiors were constructed of wood, and most fabrics were simply industrial products slightly modified for aircraft application. One of the earliest tests for flame- resistant materials was established by the FAA's predecessor, the CAA, in 1967 and consisted of applying a Bunsen burner flame to a material for 12 seconds. During that time, the material was expected to resist continuing to burn. Later, the severity of the test was boosted by increasing the flame time to 60 seconds. When it became clear that direct flammability alone was not fully representative of a cabin fire, intense radiant heat was also added to the test.

In 1986, a calamitous inflight fire resulted in fatalities due primarily to toxic gases from burning seat padding. This tragic event led to the FAA's requirement that seats in all Part 25 aircraft be covered with strongly flame-resistant, fire-blocking material. A new and more severe test was also developed, calling for an exposure to intense heat without burning for 100 minutes. In 1990, these standards were further tightened by incorporating new smoke density and toxicity standards.

Most of the major completion centers specializing in corporate aircraft interiors offer all of the newest fire- resistant materials. When KC Aviation, Duncan Aviation and others place orders for interior materials, they ensure that those products meet FAA standards. Those materials are then independently tested either inhouse or by an independent laboratory to confirm that the standards have been met.

Next, materials are tested in combinations as they would be in an actual installation, according to Ron Kitching, manager of aircraft interiors and support for KC Aviation in Dallas. "Sometimes the presence of a certain adhesive or other product in combination with another material can produce unsuspected toxicity or increased burning when subjected to testing," says Kitching. "Since each new interior is — in effect — a special order, we make certain there will be no surprises."

Present regulations require Part 25 crew members to have oxygen-only masks available for emergencies. Passenger masks which supply a mixture of ambient air and oxygen are not considered to be totally effective in cabin fires since smoke and toxic gases may be unwittingly inhaled.

The NTSB has made no recommendation on the question but has cited several instances in which passengers in toxic smoke conditions successfully inhaled through wet towels.

However dramatic the advances in producing fire-resistant materials and whatever regulations are imposed by the FAA, one simple fact remains: During an inflight fire, an alert and well-trained crew will always be the first line of defense.

Aircraft built after August 1990 are required to meet the most recent FAA flammability standards and fire-resistant materials codes. There is no requirement, however, that operators of transport category aircraft built prior to that date must replace older interior installations unless or until an aircraft in undergoing "substantial" interior renovation. No definition of "substantial" is supplied.

And, at present, no requirement mandates Part 23 aircraft, which meet lower standards for certification, to upgrade to any higher standards.

When the FAA established the current standards in 1990, it anticipated that older Part 25 certificated aircraft would gradually replace interiors. The agency also expected that by 2018, about 85 percent of all Part 25 U.S.-registered aircraft would meet the new standards.

A January 1993 General Accounting Office (GAO) study (titled "Slow Progress in Making Aircraft Cabin Interiors Fireproof") reported, however, that air carriers rarely completely replace older interiors with newer up-to-standard materials and have no plans to do so because of the cost.

By the same token, several major maintenance and completion centers specializing in Part 25 airplanes say operators of business aircraft built prior to August 1990 will sometimes "bite the bullet" and voluntarily upgrade elements of interiors (rather than the complete interior) to high flammability standards. They usually decide to have such renovation done with major refurbishing is taking place or when new avionics equipment is being installed.

Understandably, cost is a consideration.

Although the NTSB has strongly urged the FAA to establish a deadline for all Part 25 aircraft to bring interiors up to present standards, the FAA has resisted on the basis of such action being too costly for the potential benefit.

The FAA estimates that it would cost between $3 billion and $4 billion to upgrade the entire fleet. The agency has further calculated that if all Part 25 aircraft met the stricter standards, between nine and 16 lives (an average of about 12.5 lives) could be saved each year. Using a DOT- devised formula that establishes the value of a human life at $1.5 million, the potential saving if all aircraft were modified would be in the vicinity of $110 million to $115 million.

Three years ago, the Safety Board also urged that Part 23 commuters (presently held to lower standards) be required to meet the same interior flammability standards and seat restraint standards as Part 25 aircraft.

In July 1993, the FAA issued a Notice of proposed rule making (NPRM) calling for comments on a proposal to require commuter category aircraft to upgrade both seat restraint systems and flammability standards for interiors and seat cushions.

Three comments were submitted in time to meet the November 12 deadline.

The General Aviation Manufacturers Association (GAMA) requested an extension of the comment period so that its staff could examine some of the documentation justifying the need for stricter standards. A response from the European Joint Airworthiness Authority suggested a delay in issuing a rule would allow the FAA and the JAA to "harmonize" fire- resistant materials standards for commuter aircraft.

Another respondent observed that the fire-blocking materials proposed by the new standards as protection for seat cushions invariably wear out more quickly than they should.

The FAA indicated it is likely to postpone rule making for another year.

52 Fire Protection for Hotels

Since the rush of hotel fires of 1980s municipalities and insurance have encouraged the lodging industry to police itself and establish plans and fire protection systems to combat, fires. As my research continued, I noticed they cited samples of near disasters and, in some cases, actual disasters. Yet., they did not, explain how to develop a fire protection plan.

When I first became involved in fire protection and safety engineering, there was very little done to develop a comprehensive fire protection plan for hotels. Through personal research and contacts within the industry I was able to review several emergency plans that dealt with fires and fire-related emergencies.

As 1 reviewed these plans, however, I realized most of them wore too general. While they were informative and logical and offered advice, I realized more detail was needed, considering all the situations that, could be raised in a hotel fire.

For example, those plans instructed employees who discover an fire to sound the alarm. Yet, who to call, what to do, and when to do it were not clearly defined. Also, these plans instructed employees to evacuate the affected building. In an actual fire someone would certainly think of that.

However, no further information was offered beyond that single instruction. Evacuating 150 guests at 2:00 A.M. is a big order for an experienced and trained staff. In a smaller hotel, only two employees may be on duty. What then?

And what would happen to the guests and staff once they were evacuated? It might be fine to have guests stand on the sidewalk in their pajamas in Miami in June, but in Green Bay, WI, in January, it just isn't practical.

The purpose of this article is to provide guidance to security professionals and lodging industry managers charged with developing a fire protection plan. The plan must be specific and address problems that could occur in worst-case scenarios. Such as the following:

1. November 21, 1980, MGM Grand Hotel, Las Vegas, NV, 84 dead.
2. November 21, 1980, Kawaji, Japan, 44 dead.
3. December 4, 1980, Stouffer's Inn, Harrison, NY, 26 dead.
4. January 17, 1981, Inn on the Park, Toronto, Canada, 6 dead.
5. February 10, 1981, Las Vegas Hilton, Las Vegas, NV, 8 dead.

In 82 days, five hotel fires resulted in 168 deaths.

Because of the extensive coverage the MGM Grand Hotel fire received, I'd like to use it as an example of some common problems the lodging industry faced 10 years ago and, unfortunately, some properties still face today.

Consider the following facts about the MGM Grand Hotel fire:

1. Before being brought under control, there was an estimated 1 million cubic feet of fire burning at a searing 2,400° centigrade.
2. Ninety seconds after the fire erupted from the wall and ceiling, life ceased in the casino and surrounding areas.
3. At the main entrance an automobile was totally incinerated in moments.
4. There had been no formal fire drills involving the hotel staff. However, employees were instructed in the action they were to take in the event of a fire when they were hired.
5. More than half of the furniture was made of plastic that when burned gave off a toxic gas.
6. The alarm system had been rigged so that it would not ring for more than five minutes before it stopped. As a result, some guests hoard about the fire while watching Good Morning America on television.
7. The casino staff was reportedly under instructions to secure the money in the event of a disaster.
8. Guests reported fire exits were locked.
9. The MGM Grand Hotel was not equipped with an automatic sprinkler system throughout the entire property.

Building codes at the time required an area of more than 12,000 square feet to be sprinkler equipped. Although it was larger than the requirement, the casino had no sprinklers.

The International Conference on Building Officials stated the code did not apply to casinos. In fact, the code specified areas for trade shows and conventions must be sprinklered. Most lodging facilities will not call the fire department until they verify the fire and attempt to extinguish it themselves. An example of this misguided philosophy occurred in New Orleans. Fire department records show that it received a telephone call from a guest in a guest room. The front desk had been notified approximately 20 minutes earlier and had sent a security officer to investigate the cause of the fire alarm. The security officer's body was later found on the 12th floor about 10 feet from an elevator.

How do we secure a hotel or motel from fire? Perhaps we don't. We security managers and directors have responsibility to provide a safe environment for our guests and employees. We must do everything possible to upgrade and maintain fire safety equipment that saves lives and protects property. We cannot turn our backs on the life safety issue simply because a property is unable to afford state-of-the-art equipment. We must work to implement alternatives and develop an effective fire protection plan.

A fire protection plan can be set up in four phases: prevention and preparedness, reporting procedures, response and evacuation, and recovery operations.

Prevention and preparedness are the single most important element in a fire protection plan. If a well-thought-out prevention program is in place, the chances of a fire occurring are greatly reduced.

The prevention section of the fire protection plan should establish stringent standards to use during fire inspection tours of the facility. These standards must address the following:

1. A detailed fire inspection of the property should be conducted regularly. A weekly inspection is recommended.
2. Laundry equipment lint traps and filters should be cleaned regularly. A thorough cleaning on a weekly basis is recommended.
3. Range hoods, kitchen vents, and related duct work should be cleaned regularly. Hoods and filters should be removed and cleaned daily, with the related duct work cleaned at least every six months.
4. Emergency shutoff points for all fuel lines should be clearly marked, and access to then should be unimpeded.
5. In areas protected by sprinkler heads, equipment and products should be stored no closer than 24 inches to the head. To store anything closer will adversely affect the water flow and spray pattern in the event of a fire.
6. All fire alarm pull stations must be easily accessible.
7. Garbage must be removed often to prevent buildup.
8. All fire extinguishers must be checked daily, with a weekly written report on their condition.
9. All fire exits must be operational. Stairs and corridors leading to these exits must not be used for storage. Employees should be assigned to open every fire exit daily. This is especially important in areas where there is heavy snowfall. During snowstorm, snow can easily build up against the fire exit and prevent it from opening. In this scenario, someone should be assigned the task of shoveling out the fire exits.
10. Emergency telephone numbers must be prominently displayed on every telephone in the property.
11. Candle use is generally strictly controlled by local or state fire codes. In the event it is not controlled, candles should be lit only within the confines of a glass chimney.
12. Telephones in passenger elevators must be in operating condition and tested weekly.

Storing flammables and combustible products is always a problem. With the enactment of the Hazard Communication Program, the Occupational Safety and Health Administration has provided specific information outlining flammable products that may be kept in stock. The material safety data sheet is an excellent source document to use when determining which products are combustible or flammable.

Ideally, these products should be stored outside of the building that houses guests. If this is not possible, the storage area should be sprinkler equipped. One item often overlooked in the search for flammable products is the torches the engineering department uses.

Designate no-smoking areas. Attics, linen rooms, flammable storage areas, laundry rooms, utility rooms, and maid's rooms are all candidates for the no-smoking area designation. Enforce the policy.

Establish a hot work policy. Whenever the engineering department or a contractor is using a torch, welder, grinder, or other flame- or spray-producing device, mandate that a fire watch be established. The fire watch should be present during the entire time the work is going on and at least one hour after the work is completed.

Test smoke detectors, pull stations, and heat detectors annually at a minimum. Maintain a record of these inspections. Have room attendants check smoke detectors in guest rooms when cleaning. It is not uncommon for persons who are freebasing cocaine to stuff cotton into a detector or place a shower cap around the unit to prevent it from going off. Emergency lights and generators should also be checked and tested in accordance with the manufacturers specifications.

Also, review job descriptions for all employees. A job description is a good place to identify the responsibilities of each employee in the event of a fire.

On the day of the fire what the employee actually does may be significantly different from what the job description states but, in the event no one is around to assign duties, the employee can fall back on the job description.

Plan ahead. Where will the evacuated guests and employees assemble? Make the necessary arrangements ahead of fire. For example, plan to have the guests walk to a nearby cafeteria, theater, warehouse, school, or mall. In a resort or other remote facility, transport guests by bus to an evacuation point. In larger resorts and hotels with multiple buildings this may not be a problem.

Determine ahead of time who will supervise the assembly area for guests and employees. This individual is going to be on the hot seat since guests will have many difficult questions. All employees should be advised during their initial training not to cake statements regarding how the fire started or accept responsibility on behalf of the hotel. Designate one spokesperson.

Guests do not have to be disturbed by fire alarms during fire drills. It may be more practical if the alarms don't sound at that time. After all, you don't have any guarantees the fire alarm system will not be the first casualty of the fire. During these drills, have employees do what they would do during a fire. Employees should be able to explain error how they will use the fire extinguishers.

If you have an entire wing, building, or floor empty for an evening, use it for training. Have the local fire department come in and walk through the facility. The fire fighters' responses will be improved if they have walked the building and have an idea of what they will find when responding.

Develop a radio or pager call code for fire smoke. This is important if you hope to control the evacuation. Imagine: A member of the security staff could be walking through a crowded ballroom when the radio report of a fire in the trash compactor is made. What will the people in the ballroom hear? One word—FIRE. You and your staff now have a problem as 1,000 people head for the fire exits.

Establish emergency equipment cabinets at strategic location in the facility. These cabinets should be outfitted with equipment you will need during an actual fire emergency. Stock them with pressurized water, carbon dioxide, and dry chemical fire extinguishers; first-aid kits; fire blankets; and flashlights as a minimum.

Coordinate with the local fire department as to how your entrances will be set up for emergency responses. Instruct your employees that when calling the fire department, they should tell the dispatcher which entrance the fire department should use. Number or letter your entrances clockwise or counter-clockwise beginning with the first entrance the fire department will encounter as it arrives.

Ensure hallways and corridors are kept clear. They should not be used as additional storage space. The corridor must be at least 36 inches wide, the width necessary for a wheelchair to pass.

Establish a policy so guests and visitors can easily identify employees during a fire. During one fire, employees tied white cloth napkins around their upper arms. The guests then knew who the employees were in the event they needed assistance.

Place instructions in guest rooms as to what action to take during a fire. These instructions must tell guests not only the location of the fire exits but also how to survive a hotel fire.

Reporting procedures. As noted earlier, many facilities do not call the fire department until after they have confirmed the report of a fire. Minutes are lost, property damage can easily double or triple and lives may be lost during this delay. This entire concept is dangerous and outdated.

Encourage your employees to notify the fire department as noon as a fire is reported or detected.

This policy of notifying the fire department before attempting to fight, a fire has been debated within the industry for quite some time. Management worries about what the guests will think; loss prevention professionals worry about life safety and property damage. If a guest opened his or her guest room door and saw six fire fighters in full gear walking down the hall, the guest would be very concerned, too. I believe guests would rather have the fire department on the premises than left wondering about the smoke they smell. Individual properties have to wrestle with this issue; I hope they will decide to notify the fire department as soon as a fire is reported.

When planning for a fire emergency, employees should be advised exactly how they are to report a fire. Who the situation is reported to, how it is to be reported, and what the person receiving the report is to do must be carefully outlined and explained.

In addition, the sequence of notification should be listed. Again, the fire department, should he notified first. Follow up by notifying the manager on duty, security staff, in house fire brigade, and whoever else is appropriate.

The most reliable means of communication should he used when reporting a fire. Don't rely on a pager system for the initial reporting process. The person making the initial report has no way of knowing whether the person who was intended to receive the page has actually received it. Telephones or two-way radios are the best devices for reporting a fire.

The fire alarm system in the property should operate and should not be configured to switch itself off automatically after sounding for a short period. The required fire alarm signaling system should ideally be supplemented with a public- address system.

Some facilities have a recording played when the fire alarm system is activated. Other facilities rely on live- broadcasts by managers to direct guests on how to respond during a fire. Whichever method you choose, ensure the guests are given current and accurate information.

Response and evacuation. Responding to a hotel or motel fire at 1:00 pm on a Friday, with a full staff on duty, should present few problems if a well-rehearsed plan is in place.

However, the plan must also address how to respond to a fire emergency at 3:00 am on a Sunday with only a handful of employees available. It is especially critical that each employee know and be capable of carrying out his or her assigned fire emergency tasks at any time.

Do not expect your staff to respond to a fire scene alone. Even when they are checking it out, hazards are present. Professional fire fighters rarely enter a fire scene alone; they enter in teams. To expect your staff to take greater risks is absurd. During a fire emergency, life safety must be your primary concern.

When a fire is reported, sound the general fire alarm. Let it ring. After all that's why the system was installed. If you're going to silence the alarm after it rings for a couple of minutes, there's not much point in having it.

Make it a policy that once the alarm sounds, employees are dispatched to contact guests. Every guest room should be contacted by telephone and then again in person. Every guest room should be checked for guests as soon as possible after a fire alarm either by staff or rescue personnel. There have been situations where guests thought the call was a prank phone call.

Many guests can evacuate themselves given proper instruction and direction. However, the property's fire protection plan must also address how disabled guests will be evacuated.

For example, how are you going to get a person in a wheelchair down 25 flights of stairs while competing with other evacuating guests and employees for stair space? Assuming you have a plan to accomplish this, are members of the staff physically capable of handling a 200-pound person in a wheelchair?

Special equipment today makes it easy to transport wheelchair-using persons down stairs. In addition, consideration must be given to tracking rooms with disabled guests. During a response by the fire department, the manager on duty must convey these room numbers to the on-scene fire commander so the fire department can help evacuate such guests.

Fight or flight? Often the question will be, Do we fight the fire, or after evacuating the guests do we evacuate all employees as well? I don t believe that in the interest of life safety there really is a question. Most deaths in a fire result from smoke and gas inhalation.

Have the staff been trained in proper response techniques at a fire scene or are they on their own? Do all employees know the capabilities of the fire extinguishers in the building?

Most employees do not know how the units operate or, tragically, that the fire extinguishers are even present. Employees have a tendency to not see fire extinguishers because they are always there.

Train the staff in response and fire-fighting equipment use.

The Fire Safety Handbook, published by the National Fire Protection Association, lists the common types of fire extinguishers and their capabilities. The accompanying chart shows the three most common fire extinguishers found in the hotel and motel environment and their capabilities.

The approximate time of discharge column represents how long the fire extinguisher will discharge the extinguishing agent when the valve is held open. The discharge times are all one minute or less. When employees are responding to a fire scene, they must know the limits of their equipment.

During response to the fire scene, the authority of every member of the stuff must be clearly defined. Too often the person designated as the fire marshal for a building or other area is a low-ranking person in the organization. After the organization has spent money to train the fire marshal, this person may well be the best-trained individual available during a fire.

However, during an emergency, organizational structure almost always takes over. Will a senior manager take instruction from a bellman of food server? If this same senior manager is now issuing orders during an emergency, even though he or she is not properly trained, will the remaining staff listen to the bellman turned fire marshal or the senior manager?

Every organization I've been associated with dictates that when a senior manager talks, everyone listens—even the better trained fire marshal. Thus, senior managers must be involved in the planning process and the fire drills.

Evacuation must be conducted over routes employees are familiar with. Ideally, every employee should be familiar with every fire exit and evacuation route. If a real fire emergency develops and an employee who has never been on the fourth floor is pressed into service during the evacuation, a problem can quickly develop when the stairwell he or she came up fills with smoke. The employee now has one concern: How will I get out?

An in-house fire brigade should be established. When the fire alarm sounds, members of the brigade should meet at a designated emergency assembly area where emergency equipment is available.

After meeting, the team should go to the scene of the emergency. As it nears the scene, the team should be mindful of dangers associated with the situation. Prior to going through a closed door, for example, someone should feel the top and bottom of the door. If there is a temperature difference, the fire may be oxygen starved on the other side. When the door is opened, the smoldering flames will get the oxygen they need, creating a hellish inferno for the person who mistakenly opened the door.

Members of the team must remember they are not fire fighters. They must limit their activity to closing doors and limiting the extent of the fire damage. Team members are also responsible for getting responding fire fighters into the building and to the scene of the fire without endangering their own lives.

During the response the staff may also be confronted with parents attempting to get into the building to rescue children. Plan ahead for this scenario. How will you keep the parents from entering a burning building and yet assure them everything possible is being done to rescue their children? This is another reason guest rooms must be physically checked when an evacuation occurs. When coordinating with the fire department during the planning phase, you should address this problem and seek assistance.

Finally, when planning the response phase of your fire protection plan, plan on a worst-case scenario: no electricity, no water, a delayed response by the fire alarm system. The staff that responds properly to such situations will easily be capable of handling lesser problems.

Train for major fires with a small contingent of employees, coordinate with the fire department, and provide detailed instructions in the fire protection plan. Your team of employees will be able to handle most fire effectively until the fire department arrives.

Recovery operations. The fire is out, and the fire department has completed its overhaul and is wrapping up its operations. The fire emergency is over. However, the property must now engage in another fight—a fight for survival. The fire protection plan must address what will happen after the fire department goes, leaving you with an area of charred rubble the size of which depends on the seriousness of the fire.

What if guests are not allowed back into the building? Is the building safe? What are you going to do with evacuated guests who have only the clothes on their backs? The list of questions is virtually endless. Plan ahead and have sere of the answers.

First things first. Get the guests into new facilities.

When planning for this emergency you should contact other local hotels and motels and know their capacities. Arrange for transportation. Make arrangements for guests to get necessary toiletries at your expense. Spending a few dollars now may cultivate goodwill with guests who may otherwise seek damages through civil action that could prove costly.

Don't forget you have guests checking in, possibly at the same time the fire department is leaving. An updated copy of reservation information should be maintained at an off-site location and updated daily. You may have to call guests and advise them that you may not be able to accommodate their requests.

Evaluate the building as soon as possible after the fire. If it is safe and the damage is limited to one specific area, prioritize the guest rooms and public areas. Plan on getting guests into the facility as quickly as possible. Set up a priority chart using letters or numbers.

For example, "A" would represent rooms and areas that require normal cleaning and ventilation and can be ready for guests within 24 hours. "B" can represent areas and rooms that require additional work but can be made ready for guests in 48 hours without replacing furniture, walls, or other fixtures. Areas and rooms categorized as "C" can represent areas that must be painted and thoroughly cleaned and that may have sustained some minor water or smoke damage. These areas can be made ready for guests within seven days.

Finally, areas and rooms listed as "D" may require help from contractors. These are the areas where the fire started and where the majority of the damage occurred.

Developing a fire protection plan takes time and effort.

When you have taken the time and expended the effort, conducted the training, and reviewed procedures, dealing with a fire may not be devastating as you think. If the fire protection plan addresses prevention, reporting, response, and recovery, fire situations can be managed rather than reacted to.

53 Plant Safety and Fire Protection

Fire protection engineers and safety professionals frequently are asked to use their plant's existing resources to meet federal, state or local regulations. Capital constraints and lack of knowledge of specific regulatory requirements make that a reality of business life. But for emergency alerting systems, making due with existing equipment may not allow a facility to meet today's minimum regulatory requirements, yet alone ensure that it is prepared to continue operating in the future.

In one plant the existing emergency alarm system was evaluated, and it was found that the system was inadequate both for the scale of the site and for the complexity of its operations. The emergency alarm was sounded through a steam fire whistle activated from the powerhouse. Despite its intended use to call the shift and volunteer emergency response teams, and to warn all employees, this whistle was not capable of being heard in all areas of the refinery.

Some 140 radio frequencies were in use at the refinery, with the yard frequency designated for emergency management and tactical communication. The emergency action plan called for partial or total site evacuation orders to be given through the yard frequency.

Some process units had individual alarm systems, activated by a push-button, that broadcast evacuation warnings in case of an emergency. These systems, some of which did not meet the area electrical classification requirements, were installed for major maintenance repair work (turnarounds) to warn contract employees.

Portable air horns were used by designated floor wardens to initiate emergency building evacuations.

The refinery had 1,800 telephone numbers, one of which was designated as the primary method for employees to report emergencies. During off-hour emergencies, and at the request of the operations shift superintendent, security called out key personnel. Individuals not responding to the phone calls were paged.

There are several OSHA regulations setting requirements for industrial emergency alerting systems that were net through this project:

1. Employee Emergency Plans and Fire Prevention Plans requires a written emergency action plan that has to be made available to employees.

2. Since the alerting system was going to be used for purposes other than just to issue employee evacuation orders, a distinct, signal for each purpose needed to be used. Under paragraph (q) Emergency Response to Hazardous Substances Releases of OSHA's HAZWOPER regulation, such purposes include calling fire brigade members or notifying employees for emergency response activities.
3. Paragraph (n) Emergency Planning and Response requires that the refinery emergency action plan meets OSHA 1910.38. In addition, paragraph (h) of the PSM regulation requires that all affected contractor employers be informed of the applicable provisions of the emergency action plan. This is one of the areas most difficult to meet without significant effort and business interruptions when you have a contract work force in excess of about 500 persons, possibly working in different areas, and you use a steam whistle for emergency warning without additional means to provide specific instruction in a timely manner as to the nature of the emergency and how to proceed.
4. To satisfy the requirements of these three previously mentioned regulations, the emergency action plan had to include an emergency alerting system that complied with this OSHA regulation.

We know that the adequacy of an emergency action plan is in direct proportion to how practical, it is to implement the action specific in the plan. In theory, all actions may seem possible to implement. In reality, during major emergencies, accurate and timely emergency warnings to all affected people are among the most important factor to successfully protect lives, contain property damage and minimize community impact.

The refinery's steam fire whistle clearly was not adequate means to notify all employees, emergency responders, contractors and visitors of whatever one might need to know in case of an emergency, in a timely and reliable manner.

Our task was to design and install, a state-of-the-art emergency alerting system that met all pertinent regulation as well as CITGO's needs, caused minimum disruption to refinery operations, and could be installed within a year's time at a reasonable cost.

To help ensure that the alarm was effective, process area ambient noise level contour plots prepared for our hearing conservation program were evaluated. This was done to identify all high noise areas within the refinery. An acmustic computer model was then used to predict speaker systems sound coverage and to identify the optimum location of each speaker station to ensure clear, intelligible voice messages that could lie heard throughout the protected area.

The computer model uses a 10-decibel differential above ambient noise for the entire acmustic coverage area, and takes into consideration factors such as: hemispherical wave divergence, atmospheric absorption of sound, wind effects, ground cover absorption, and barrier attenuation effects (such as those posed by buildings, atmospheric tanks, and above-ground process equipment). Figure 1 illustrates a sample of an acmustic coverage analysis.

The proposed outdoor speaker locations were then evaluated for suitability regarding safety, operation, installation and maintenance factors. This resulted in some proposed speakers' locations being changed. Finally, to verify that the speakers provided adequate coverage for tone and voice and that the message would be intelligible, field test measurements were conducted for each proposed location.

Tests were conducted using wireless directional speakers that provided high-intensity sound beams to overcome high noise areas. These field tests confirmed the coverage predicted by the acmustic model.

Each outdoor speaker station in the process area consisted of a pole with eight to twelve 100 watt speakers mounted on it.

The controllers for the speaker stations incorporate an amplifier with a two way wireless radio frequency linked to the primary and backup central stations. The refinery dock areas needed a special configuration of the 100-watt speakers to direct sound toward the operations area of the dock and to eliminate sound propagation beyond the refinery property.

A total of 23 high-power speaker stations were installed.

These stations are capable of broadcasting live public address messages, prerecorded voice messages and emergency alerting tones. They provide monitoring of their operational status and operate on battery in the event of an AC power loss.

The link between each speaker station and the emergency alerting system central control unit (CCU) was achieved through remote controllers (RCs) and radio frequency communication. The RCs consist of a single-board controller, radio and antenna, battery and battery charger, transformer and surge suppressor. The units are contained in a stainless steel NEMA-4 enclosure.

Manual Alarm Activation

The design was based on the premise that every person in a process area available means to initiate an emergency alarm through the use of push-buttons. This condition was critical for a prompt emergency notification because telephones are not available in all outdoor locations and not everyone had radios to initiate an alarm. The push-buttons were to be located in:

1. Push-buttons were located inside process units so that no one had to travel more than 100 feet in order to reach a push-button. The location had to be visible from each evacuation route covered by that push-button, well-lit, and clear so that a person would not have to navigate through equipment, piping, instrumentation or other possible obstructions to reach the push-button.
2. Each process operator's shelter would be provided with an emergency push-button, visible and accessible from any shelter entrance and able to be activated without leaving the shelter.
3. Emergency push-buttons would be installed at each one of the refinery unit operating consoles, located inside the three existing control rooms.

A total of 85 wireless pushbuttons were installed in outdoor locations throughout the refinery, using the latest technology of reliable spread spectrum radio transmission.

The license-free 900 megahertz (MHz) radio frequency uses the same technology as cellular telephones and works well in the refinery because the signals actually reflect off metal in the process units. To provide higher reliability, the system utilizes dual-frequency techniques and diversity antennae that virtually eliminate signal canceling. The pushbuttons are lithium battery-operated, designed for use in Class 1, Division 2 environments. The batteries have 5 to

10 years life span and provide a 30-day notification of transmitter low-battery condition. The wireless push-buttons continually report into their associated receiver, and generate audible and visible alarms at the central control unit in case of failure.

The wireless design allows for future expansions, provides great flexibility for locating the push-buttons and reduces the installation cost because AC power is not required. (A wireless push-button unit and a remote controller are shown in Figure 2). Another important benefit was that the installation did not disrupt the units' operations.

In addition to the 85 wireless push-buttons, 25 conventional hard-wired push-buttons were installed in control rooms and operator's shelters.

Although the outdoor speaker stations' acmustic range was designed to cover the entire refinery, it was understood that inside buildings, especially the sealed control sound level of the alert tone and the intelligibility of the voice messages would be reduced. For this reason, speakers were installed inside buildings and control rooms to ensure 100 percent emergency alert coverage.

The number, placement and wattage rating of indoor speakers and amplifiers were based on the specific conditions of each location. Factors considered included the square footage of the building and layout, number of floors, and the existing background noise level. Indoor speakers had to be capable of broadcasting live public address messages, prerecorded voice messages and emergency alerting tones.

The indoor speakers are connected to wireless microprocessor- based remote controllers (RCs) operating two-way wireless frequency linked to the primary and backup central stations, and provide status monitoring of its operation. Both the indoor speaker stations and the remote controller can operate on battery in the event of an AC power failure.

Wireless Radio Communication

The emergency alerting system uses the refinery's 900 MHz trunked radio system to link the central control unit (CCU) and the remote controllers.

To transmit, data via radio frequency, the system uses continuous phase frequency shift keying and data error detection accomplished through extended parity checks for every 8-bit block of data within a message.

A multiple access hierarchy allows the emergency alerting system activation and monitoring to coexist with other refinery radio transmissions without the loss of data.

Battery backup for the entire system allows operation in the event of a main power failure.

With the old system, vital time was required to call key emergency responders and support personnel during emergencies. Individuals or groups were called on pagers, some of which had voice capability with a very limited coverage area (approximately 10 miles), while other pagers had just numeric display and about 40-mile coverage. Due to systems incompatibilities and limitations, some key personnel were assigned two pagers.

The new system employs approximately 300 alphanumeric pagers which can identify the type of emergency and its location. Now, when an emergency alert is activated by phone, radio or push-button, the on-shift incident commander (shift supervisor), tactical chief (shift fire-safety inspector) and security sergeant pagers are automatically activated with a first alarm emergency notification. If requested by the incident commander or his designee, the central control station operator can activate a second alarm for the on-shift emergency response teams.

A third alarm is available for all safety and industrial hygiene personnel and key maintenance supervision, while a fourth alarm is used for all volunteer emergency responders and refinery personnel with defined responsibilities in the emergency action plan.

Individual and group messages can also be sent to the refinery rescue and hazmat team members. Approximately 65 percent of the pagers have a 40-mile area coverage. The remaining pagers have Gulf Coast coverage through a mobile communications network for safety and industrial hygiene, operations, maintenance and technical services management and supervision, and refinery senior management personnel. The new alphanumeric pagers retained the previous individual pager numbers and also are used for non-emergency calls.

Control Stations

Primary and backup emergency alerting system control stations were specified, each one equipped with a central control unit (CCU) which is the microprocessor command station for the system. The CCU is connected to a 486SX computer equipped with a VGA monitor.

This configuration provides two totally independent stations (main located at the refinery powerhouse control room and the backup installed in the security office) capable of managing the system activation, continuously monitoring the status of the main system components and the radio communication link. The system also has a live public address capability. The alarm system software used on the system is easy to operate and requires no more than 2 hours for qualifying its operators.

After a couple months of training all control station operators (powerhouse console operators and shift security sergeants), performing field testing and making minor final adjustments, the plant emergency alerting system was commissioned.

54 Safety Training

A new era in worker safety and health has arrived in the past two decades, and the need for training is being addressed by regulatory and legislative bodies. The responsibility for safety training typically falls under the jurisdiction of the safety and health professional. Whether members of the safety department actually conduct the classes or merely ensure they are conducted depends on the individual organization. Organizations can no longer simply provide a training session, however; the emphasis must be on workers demonstrating satisfactory understanding of how to perform various work procedures using safety equipment, proper guarding, and proper maintenance. With this in mind, it is critical for the safety and health professional to work closely with all levels of management and employees to ensure that training is adequate, accurate, and timely.

Management from top executives to supervisors must recognize the importance of training and support training efforts if these programs are to have any validity. When employees believe that safety training is not important to their supervisors, it will not be important to them.

There are several ways for the safety and health professional to identify the training needs of the organization. The professional can work with company legal counsel and others to stay current with all regulations mandating specific safety requirements. Accident and injury records together with specific investigation reports can reveal areas of weakness or failure in training. Although this approach looks at the past, it can be turned into a proactive approach designed to prevent similar incidents from occurring.

According to Heinrich's theory, past events forecast the possibility of future and potentially more severe incidents.

Other ways the safety and health professional can determine training needs include workplace surveys (safety audits), regularly scheduled inspections, and employee opinion surveys. These are preventive methods used to identify training needs so that programs can be established before more accidents occur.

Safety and health professionals must keep in mind that training programs do not eliminate a company's obligation to provide safe working conditions or to remove unsafe materials from the workplace. Training, however, will ensure that employees know how to do their jobs using the proper methods. Reports have strongly suggested that the majority of accidents are caused by workers' unsafe practices and errors rather than by equipment failures or poor engineering. At least in part, training can change undesired behavior and/or reinforce the desired behavior.

Training, unfortunately, is not the answer to all work- related problems. It is only one method of influencing human behavior. Before a training program is developed, management should thoroughly evaluate all problems and expectations related to the operation to determine if training will accomplish the desired goals. It must be clearly understood by everyone that training can be used only to correct skill deficiencies or to improve knowledge. Most training programs cannot correct poor engineering, poor worker attitudes, or substitute for good safety engineering and enforcement.

Training Policy

In many organizations, written policy statements establish the purpose, goals, and procedures for such areas as the company mission, hiring practices, purchasing, and so on. A written statement should be developed to cover the organization's policy on training. This may be incorporated into an overall policy statement or into separate policies for each issue.

Whichever method management, selects, it should keep certain considerations in mind. The first concern is that all policies must be written to work together to achieve the ultimate company objective. Too often policies are developed separately and tend to conflict with one another. When this occurs, supervisors and workers become confused and may show inconsistent compliance. Thus, management must make sure that when several policy statements are used, they do not contradict one another.

A second consideration is the development of training schedules for all employees. A common error, found in many organizations, is to focus on safety training for workers and possibly supervisors and neglect to provide supportive training for management. This oversight can develop into a "conflict of interest" situation because the management team is unaware of the activities of the safety program. When managers understand the organization's safety efforts, they find it easier to support them.

Therefore, the organization's training policy should define training programs for all levels of management and employees. The importance of this policy cannot be underestimated. Many of the difficulties companies experience in trying to provide safety training are a result of communication problems between different levels of the management team. Likewise, many of these same problems could be resolved by training managers in such areas as their legal obligations, their safety mission, savings potential of safety training, and other concerns.

A strong safety policy will avoid pitting the safety organization against management over training issues. The policy should identify the type of training to be provided for workers, line supervisors, and management personnel.

Elements of the program, such as the type of training methods used and the location of the training, may also be specified. A well-planned training program should never interfere with production schedules. The policy statement can help to reinforce the principle that safety training is good business and can actually improve production rather than interfere with it as some managers may believe.

Planning the training program in advance will help to prevent any friction that may otherwise occur between departments. When management staff is guided by a policy statement, the training program becomes part of the company's overall operational standards and will be incorporated into the schedule rather than regarded as an interruption.

Safety Training Needs

A safety training program in needed under the following conditions:

1. For nil now or reassigned (transferred) employees
2. When now equipment or processes are introduced
3. Whenever procedures are revised or updated
4. When new information is available or required
5. When employee performance needs improvement, or to provide employee growth
6. When employee interest in safety and efficiency needs a boost.

Many organizations mistakenly assume that transferred employees will not require orientation to their new assignments because they already know the operations. This assumption has been responsible for many worker injuries. In many instances, a transferring worker requires almost the same type of orientation as does a newly hired employee. Information should be presented on the location of safety equipment, Material Safety Data Sheets, MSDS files, proper accident and first aid reporting procedures, evacuation procedures, and other items differ from workstation to workstation. For transferred employees to work safely and efficiently, management must orient them to the new environment.

Certain symptoms will indicate to the alert safety and health professional that training is needed. These symptoms include:

7. Proportionately higher accidents or injury rates than other companies that do the same type of work
8. Increasing accidents rates and/or insurance rates
9. High employee turnover
10. Excessive waste and/or scrap materials from operations
11. Recent company expansion or procedural changes
12. Increased or high levels of sick days
13. Changing regulatory requirements
14. Need to upgrade or update employees' knowledge and/or responsibilities
15. Poor job satisfaction.

Organizations that have had great success in developing safe working environments understand the importance of the supervisor's role. Therefore, they focus considerable safety training on first-line supervisors. This does not eliminate the need to train all managers and employees but provides sound leaders and good management in the safety effort.

It is important to note that the training program for supervisors includes human relations and communications skills. The supervisor is the link between upper management and the workforce. With proper training, supervisors can provide guidance, set a positive example, resolve problems, and identify the concerns of employees. Unless an organization provides the necessary leadership training, it cannot expect to have effective leaders.

Program Objectives

The first and most important step in developing a training program is to identify those who need instruction and to establish program objectives. The objectives must clearly state what students are expected to gain from the program.

In order to do this, objectives must be measurable. Because employee attitudes and awareness generally cannot be measured, they are not considered proper training objectives. Preparing course objectives requires training and practice. There is too much time and money at stake to allow a "hit or miss" approach. If students are to be taught a particular skill, the objectives must state the measurable level of performance desired. By detailing the program objectives, instructors can develop a training session to meet the company's established goals. Objectives also serve as a means of evaluating the training's effectiveness.

Weak or vague objectives often result in poor or unnecessary training programs. Excessively high objectives may overburden the students with unless information. With good course objectives, the students can focus on the essential requirements of the program.

Prior to stating the objectives, management should review workers' responsibilities. The job description or Job Safety Analysis (JSA) may provide the necessary information.

Supervisor observations and performance testing can also identify areas of discrepancies. For example, a supervisor's job description may include the responsibility of training new workers in proper work methods. Thus, supervisors will need to receive education and experience in training skills. Organizations can make a costly mistake by assuming that everyone who knows how to perform a job also knows how to teach it. Teaching is a skill that requires not only training but practice and performance evaluations. A poor trainer can be the source of considerable problems in safety, quality, and production. As a result, the cost of not training the supervisors would most likely be far greater than the cost of a "train the trainer" program.

As safety and health issues become more complex, organizations must work hard to keep up with current requirements. Therefore, the objectives for training should be reviewed at least annually to ensure that they meet organizational needs. Many organizations are using the team approach to evaluate company training needs. The team may be comprised of members from the workforce, line supervision, safety department, maintenance, and management.

Supervisors and managers must demonstrate their support and approval of the training objectives. Workers are quick to detect when management support of a program is lacking. They will generally follow their supervisor's lead in either disregarding or adhering to safety procedures.

A lesson plan is often referred to as a blueprint for training. Trainers can choose from several types of lesson plans. The selections should be based on organization requirements and instructor preferences. However, once the decision is made, all the company lesson plans should be created in the same format. This approach will make it easier to review the plans periodically. The lesson plan is designed to help the instructor:

1. Present material in the proper order
2. Emphasize material in relation to its importance
3. Avoid omission of essential material
4. Keep classes on schedule
5. Provide trainee participation
6. Increase trainees' confidence, especially new employees.

The order of information within the lesson plan may vary with the style of each instructor, but should contain the following:

1. Title: Should clearly identify the subject matter
2. Objectives:
   a. Indicate the requirements for successful completion
   b. Establish level for course material
   c. Stimulate thinking about the subject
3. Training aids: Identify materials used in course presentation such as videos, charts, equipment, or tools
4. Introduction:
   a. Establish the scope of the subject
   b. Demonstrate the value of the information to the student
   c. Focus the student's thought on the subject material
5. Presentation:
   a. Provide the plan of action in a logical sequence
   b. Describe the method of teaching to be used, such as lecture, demonstration, class discussion, or any combination of these
   c. Provide direction for the instructor activity, for example, information to be written on chalkboard or use of visual aid
6. Application: Indicate how this material or information will be utilized by the students
7. Summary:
   a. Restate the main points of session
   b. Tie up any loose ends or confusing issues
   c. Allow for questions and answers
8. Test (if applicable): Determine the level of understanding and/or performance to ensure the course objectives have been net
9. Assignment: Provide references or instructions for preparation of future classes
10. Type of certificate, diploma, etc., to be given upon completion.

Training Methods

Instructors can select one of several different training methods when preparing a program. Each method has strengths and weaknesses. The technique selected should be determined by the objectives to be net, type of student participation, time allocated, facility being used, and equipment available. Everyone learns at different speeds, and through different methods. Instructors must have the training and teaching skills to address these elements of human behavior. The following are the most common types of training techniques used in industry:

1. On-the-job training (OJT)
2. Group techniques
3. Conference method
4. Individual techniques
5. Computer-assisted training (interactive)

55 Flame Retardant

Halogenated flame retardant (FR) additives have been used in engineering plastics for many yours, but arc coning under closer scrutiny because of purported health hazards. While regulatory action regarding these ubiquitous chemicals is in check in the U.S., European regulations appear to be gaining more support. Thus manufacturers serving worldwide markets for electrical equipment are seeking alternatives.

Generally less expensive than other inorganic, organic, or mineral FRs, halogenated FRs include brominated diphenylethers, brominated carbonate oligomer, brominated polystyrene, and brominated phthalimide. They work by a vapor phase mechanism; that is, during combustion they produce a gas that impedes the spread of a flame. But gases released by the halogens during combustion can be potentially harmful and corrosive. As a result, brominated FRs are under suspicion in the environmental community. Specifically, brominated diphenylether usage is being steadily reduced because of its potential to form highly toxic dioxins and furans during combustion — byproducts classified as carcinogens.

Replacing halogenated FRs in tough. Researchers are hampered by a lack of alternatives that are an effective, efficient, and economical. Also, nonhalogenated FRs do not work with as many different engineering plastics.

Two nonhalogenated FRs used with engineering plastics are triarylphosphate and red phosphorous. Triarylphosphate is used with polyphenylene oxide and polycarbonate/ABS blends. Red phosphorous is used with glass fiber reinforced nylon 6/6.

Red phosphorous and triarylphosphate do not employ a vapor phase mechanism. Instead, they form a layer of char and glassy polyphosphates on the surface of a molded part exposed to a flame. The protective layer stops fire spread by shielding the molding from oxygen and heat. Because of the FR mechanism by which red phosphorous and triarylphosphate operate, there is low smoke density and no corrosive gases formed during combustion.

Although electrical equipment designers select flame-retardant materials from a palette of engineering plastics, they often specify glass fiber reinforced (GR) nylon 6/6 for its high dielectric strength, mechanical properties, heat resistance, and tracking resistance. Normally, GR nylon 6/6 is rated as slow burning. Adding rod phosphorus yields a material that has a UL 94 flammability rating of V-0 at 1/32, 1/6, and 1/8 in. thicknesses.

Red phosphorous normally imparts a rod color to moldings. However, black FR GR nylon 6/6 grades are available. The red phosphorous FR GR nylon 6/6 grades also contain a stabilizer that prevents the phosphorous from reacting with moisture and forming phosphates on the molding surface, which would impair electrical properties.

Flame retardant GR nylon 6/6 grades are available with glass fiber contents ranging from 25 to 50%. The flexural modulus of these grades range from about 1 to 2 million psi and their tensile yield strength ranges from about 20,000 to 30,000 psi. A small amount of red phosphorous is an effective FR in these materials, so mechanical properties are virtually unaffected.

One product that takes advantage of red phosphorous is the new Westinghouse Pow-R-Line 3 "0" low-voltage panelboard from Cutler-Hammer Power Distribution Assemblies Division. The unit has four bus bar insulator supports injection molded of Ultramid A3X2G5, a nonhalogenated FR GR nylon 6/6. Supports are fastened with screws to the panelboard. Three bus bars are fastened with screws to the nylon insulator supports and circuit breakers are bolted to the bus bars. Thus, the nylon parts are designed to hold the chassis together and to support the weight of the components that are attached to them.

Another nylon 6/6 property, tracking resistance, makes it a natural choice for electrical applications. Tracking refers to the tendency of plastics to form conductive carbon tracks along those surfaces subjected to voltage stresses, such as arcing electrical sources. The comparative tracking index (CTI), per UL 746A, for GR FR nylon 6/6 is 500+, while typical halogen-based FR GR thermoplastic polyesters have CTIs of only 185 to 260.

High tracking resistance plays an important part in the new Safety Switch F Series from Square D Company. The safety switch has an arc suppresser housing molded of Ultramid A3X2G5, which is directly exposed to arcs produced when the switch opens. In tests conducted by Square D, the nylon housing successfully withstood exposure to 244A, 600V for 50 cycles with 10 secs between each operation. The comparative tracking index (ASTM D3638) of A3X2G5 is 500 V.

Impact modified FR GR nylon 6/6 grades are also available.

These materials have 50 to 75% more impact strength than standard FR GR nylon 6/6 grades.

High impact strength is intrinsic to the function of bus bar insulators in Cutler-Hammer's 480-volt switchgear for industrial applications. Designers use Ultramid A3XZG5, a FR GR impact modified nylon 6/6 resin. Previously, a glass fiber reinforced thermoset polyester was used to mold the supports for the Westinghouse-type DSII 480-volt switchgear.

During a short circuit, a high, sudden mechanical stress is put on the insulators, which, if cracked, would allow current to arc between phases of the switchgear or from phase to ground. In tests conducted by Cutler-Hammer per UL 1558 requirements, the switchgear's copper conductors were tied together to short circuit them, but nylon insulators were able to withstand a 200,000-A fault current, 4 cycles at 645 V, without breakage.

Typical heat deflection under load (HDUL) for 25% GR FR nylon is about 480 degrees Fahrenheit at 264 psi. This material has a continuous use temperature value of 430°F for a few hours of use and 250°F for 5,000 hr continuous use.

Red phosphorous also is an effective FR for GR nylon 6/6T, which is a more heat resistant and dimensionally stable nylon. Ultramid TKR 4365 G5, which as a UL 94 flammability rating of V-0 at. 1/8, 1/16, and 1/32 in. thicknesses, maintains its mechanical properties at higher temperatures than nylon 6/6 resins. Its maximum continuous use temperature is 320°F for 5,000 hr.

FR nylon grades that do not use either halogen or phosphorous FRs are available with UL 94 flammability ratings of V-0 at 1/32, 1/16, and 1/8 in. thicknesses. These FR nylons use either organic nitrogen-based FRs or high contents of FR mineral fillers.

Nonhalogenated and phosphorous-free FR nylons have excellent arc tracking resistance (CTI values are 600) and good dielectric properties. Typical uses include terminal blocks connectors, circuit breaker housings, and relays.

56 Fire Brigades

Building and fire codes mandate many fire protection features, but no laws require a plant to have a fire brigade. Local officials and insurance carriers may encourage or even pressure plant personnel to establish a brigade, but the final decision almost always rests with plant management.

However, if the decision is to establish a fire brigade, OSHA requirements on how it is to be organized, trained, and equipped must be followed.

Having firefighting personnel available quickly can make the difference between a minor fire and a major disaster. What happens in the first few moments when a fire is still small and controllable is critical.

Organizing, training, and equipping a fire brigade is unquestionably expensive and time consuming. Persönal protective gear alone (self-contained breathing apparatus, coat, pants, boots, helmet, and gloves) costs about $1500 a person. Recurring expenses also must be considered. For example, if wages average $15 an hour, a 1 hr monthly training session involving 10 workers and a trainer costs about $2000 a year. This estimate does not consider the trainer's preparation time, any premium pay, firefighting material used during practice, and miscellaneous items. When the cost of a recognition dinner, replacement equipment, and other incidental expenses are added in, a fire brigade is expensive.

Despite the cost, facilities of all types, including countless manufacturing plants, feel a fire brigade is a wise use of corporate assets. This belief is often reinforced when observing the fire damage and down time prevented or minimized by a fire brigade's efforts.

If the quick response of trained firefighters is important to the plant, the nest question to address is the level of firefighting expected of the brigade. OSHA regulations divide fire brigades into two categories: those that combat incipient stage fires and those that fight interior structural fires. The distinction is important, not only from a compliance standpoint, but also because of training and equipment considerations.

Brigade organization depends on the individual plant situation. If the fire hazards are relatively low and a good fire department is within a reasonable response distance, an incipient brigade will probably suffice. On the other hand, if the operation is hazardous and a good public fire department is not readily available, a structural brigade is a wiser choice.

In any event, one of these two basic types must be designated in writing and the minimum requirements met for the type of brigade that is established.

OSHA's fire brigade requirements are summarized in the table. The complete regulation is found in 29 CFR 1910.156. In general, members of an incipient fire brigade use only portable fire extinguishers or small hoses (1½ in. diameter or less) and do not have to have breathing apparatus or protective clothing. Members of a structural brigade must have protective gear and be trained to use larger fire hose and more complicated firefighting equipment, depending on the plant situation.

Because every plant is different, the makeup of each fire brigade varies significantly. However, almost without exception, among the first groups to consider when staffing the brigade is maintenance. Maintenance personnel are familiar with the overall plant operation and layout and can quickly grasp the implications of an emergency. Electrical shutdowns, which occur during most firefighting emergencies, are likely to be performed by that group anyway. In addition, maintenance workers usually are able to attend training sessions and respond to emergencies with minimum disruption to normal plant activities.

Plant protection personnel may be another manpower source, although the availability of the security staff may be limited if plant entrances must be guarded or if other duties interfere with fire emergency response. Employees involved in outside volunteer fire departments, on medical teams, or with other community emergency organizations also are candidates for the fire brigade.

Because firefighting is very stressful, the physical condition of brigade members is also significant. OSHA requires members of structural brigades to undergo medical screening, and many employers require medical exams for all its brigade members.

The volunteer nature of fire brigade membership may create problems. Some plants make participation in the fire brigade a condition of employment for certain jobs. However, such a stipulation can become a thorny issue if it has not previously been part of the union contract or past practice. Such conflicts are prevented by making membership in the fire brigade voluntary. Generally, management encourages participation by providing special recognition to members and appealing to employees' interest in preserving their workplace. The volunteer approach means that, participation may rise and fall with employees' attitudes toward their company and with various outside conditions.

For most people, job satisfaction is tied closely to recognition by management and peers. Participation in the fire brigade is no exception. To that end, many plants recognize fire brigade members, and sometimes their spouses as well, with dinners, outings, or other special rewards.

Such events should include appropriate management involvement. Brigade participation may also be encouraged by providing special articles of clothing (shirts, helmets, or caps), favored parking spaces, achievement awards, and publicity in facility and community periodicals.

Leadership and management commitment must be strong. As with any other organization, fire brigades swindle and die more often from lack of leadership than any other reason. The brigade leader should be a respected person in the plant organization.

The number of people in the brigade again depends on the situation. Size may be limited by the number of available people, especially on off-shifts. However, lack of personnel should not deter brigade formation. Having a few employees who know what to do in an emergency is better than having none at all.

The U.S. Nuclear Regulatory Commission requires a crew of at least five trained persons be available at all times to fight fires in nuclear power plants. This number includes a leader and four firefighters to handle hoses and perform other fire-fighting work. Nuclear plants are typically remote from a public fire department. However, they tend to be well- equipped with automatic sprinklers and other protective equipment.

A five-member crew appears to be a good minimum for most plants. Having too many people on the brigade can be almost as bad as having too few. This situation makes effective training and fire-fighting difficult. Except in unusual circumstances, a brigade of more than 10 or 12 members per shift may be too large. Too many members reduce the team spirit often generated when membership is restricted to those with genuine interest and participation.

In multiple-shift operations, keeping the brigade roster current is a sizable task. If employees change shifts routinely, or are involved in frequent layoffs and callbacks, some shifts may be understaffed and others overloaded. In addition, vacations, illnesses, layoffs, and other conditions that could reduce the number of available personnel should always be considered.

Training the Brigade

The time and effort involved in training the fire brigade is a major investment, OSHA regulations require annual training sessions for incipient brigades and quarterly training for structural brigades. Most firefighting professionals view this level of training the bare minimum for developing teamwork and polishing skills. A better schedule is quarterly for the incipient brigade and monthly for the structural brigade. Each plant has a unique mix of hazards, outside influences, and experiences that influence the training program. People cannot work together effectively in an emergency if they have not practiced together. A good brigade team is a lot like a good basketball team.

Fire brigade training is no different from any other type of training. Effective training need not be elaborate. Fire brigade members usually do not need to develop the sophisticated firefighting skills of a public fire fighter. Training should concentrate on potential fire scenarios at the plant. Sessions should consist classroom instruction and actual use of plant's firefighting equipment.

Students learn more effectively by doing than through lectures or audio/visuals. Regular, recurring, hands-on training with hoses, extinguishers, and self-contained breathing apparatus Is necessary.

The length of a training session should be flexible. A good drill requiring the use and restoration of equipment may take a couple of hours. A classroom session on a recent change or current problem may last less than 30 minutes. Frequency and regularity are more important than length. Too often a trainer feels obligated to fill an hour, even though the material, and the audience are exhausted much sooner.

Good fire brigade training materials such as booklets, manuals, slides, films, and videos are available from many sources (see accompanying section, "A Guide to Training Materials"). Some effort is often needed to track down and obtain materials pertinent to the facility, and some areas need covering for which there are no suitable prepared materials. However, in most cases, a homemade session dealing with the plant's actual situation is more valuable to brigade members than polished, generic materials.

Preplanning the firefighting of particular fire scenario is an effective training session. Practicing the preplan reinforces it.

Training assistance is also available from outside sources, including the public fire service and many private firms. The fraternity created by firefighting generates more mutual support between organizations than is found in most other disciplines.

Because there is usually some turnover in brigade membership, preparing a packet of material for new members is a good idea. Subsequent training should build on and reinforce this instruction base. Elaborate brigade manuals ignored after their initial viewing end up gathering dust in the bottom of a drawer. Handouts at meetings, with a master reference notebook maintained by the brigade leader, is often a better approach.

Implementing the Brigade

Even if a private fire brigade has been established, public firefighters are in charge when they arrive. This is normally a statutory provision covering emergency situations and is not subject to negotiation.

Because public firefighters are in command, they must know any unique firefighting conditions at the plant. The two organisation should build a working relationship that allows the public officer to seek the counsel of the plant fire brigade about special problem. Fire brigade personnel, particularly the leader(s), should be permitted and encouraged to maintain frequent contact with the public fire service meeting might involve familiarization tours of the plant or visits to the fire station to discuss items of mutual concern.

Developing and maintaining an effective private fire brigade is not a responsibility to be treated lightly. It is a critical part of plant operations. If a brigade is justified. It should be well-organized, trained, and equipped. The brigade must meet OSHA requirements and should provide a significant increase in the plant's fire protection.

Although it is usually staffed by employees with other full-time duties, a fire brigade is not cost free. Management support and firefighting leadership are imperative if interest and participation in the brigade are to be sustained. A good brigade may save facilities and jobs during the first crucial moments of a fire, when it is still small.

A Checklist for Evaluating the Fire Brigade

Answering yes to the following questions is an indication of a fire brigade in good condition.

1. Does plant management actively encourage the brigade?
2. Does a written document specify which type of fire the brigade will fight: incipient or structural?
3. Is the brigade organization outlined in writing?

4. Is effective planning for staffing, training, and equipping the brigade done by those responsible for this area in the plant management structure?
5. Are adequately trained brigade personnel available on off-shifts, when temporary layoffs occur, and during heavy vacation periods?
6. Is the brigade leader effective during an emergency?
7. Have preplans for dealing with a fire in the major hazard areas of the plant been developed and practiced?
8. Are training sessions conducted regularly without lapses because of temporary problems or other demands?
9. Has brigade training given members a good understanding of plant protection systems such as fire pumps, sprinklers, and other equipment that is part of the firefighting effort?
10. Has the brigade been taught how to deal safely with the hazardous materials at the plant?
11. Has the brigade been taught how to minimize damage and salvage equipment and materials after a fire?
12. Do brigade members know where all firefighting equipment is located and how to use it?
13. Is sufficient protective clothing and breathing apparatus available? Does it fit members properly? Is it in good condition?
14. Are all members of the brigade physically able to handle their firefighting duties?
15. Has brigade training considered adverse conditions such as major storms, flooding, and loss of utilities?
16. Do members of the public fire department periodically (at least annually) tour the plant?

57 Loss Control

A workable definition of hazard is any existing or potential condition in the workplace that, by itself or by interacting with other variables, can result in deaths, injuries, property damage, and other losses. This definition carries with it two significant points.

1. A condition does not have to exist at the moment to be classified as a hazard. When the total hazard situation is being evaluated, potentially hazardous conditions must be considered.
2. Hazards may result not only from independent failure of workplace components but also from one workplace component acting upon or influencing another. For instance, if gasoline or another highly flammable substance comes in contact with sulfuric acid, the reaction created by the two substances produces both toxic vapors and sufficient heat for combustion.

Hazards are generally grouped into two broad categories: those dealing with safety and injuries and those dealing with health and fitness. However, hazards that involve property and environmental damage must also be considered.

Effects of Hazards on the Work Process. In a well balanced operation, workers, equipment, and materials are brought together in the work environment to produce a product or to perform a service. When operations go smoothly and time is used efficiently and effectively, production is at its highest.

When an accident interrupts an operation, it sets in motion a different chain of events and carries its own price tag. An accident increases the time needed to complete the job, reduces the efficiency and effectiveness of the operation, and raises production costs. If the accident results in injury, materials waste, equipment damage, or other property loss, there is a further increase in operational and hidden costs and a decrease in effectiveness. These cost factors are discussed in Chapter 12, Accident Investigation, Analysis, and Costs.

Safety Management and Productivity Improvement

The process of identifying and eliminating or controlling hazards in the workplace is one way of making the best use of human, financial, technological, and physical resources. Optimizing these resources results in higher productivity. For purposes here, productivity is defined as producing more output with a given level of input resources.

Less control, like productivity, quality personal relations, is a strategic process. To be effective, it must be integrated into the day to-day activities and management systems of the organization end must become institutionalized, an operating norm and a strategic part of the organization culture.

There are other similarities between efforts aimed at hazard control and productivity improvement. To achieve both objectives, an organization must intelligently manage its financial and human resources and use the most appropriate technology. It must illustrate innovative, enlightened, and efficient use of its facilities, equipment, materials, and workforce, and have a trained, educated, and skilled workforce.

An accident interrupts the production process. It not only increases the time needed to complete a production task, it may also reduce the efficiency and effectiveness of the overall operation and increase production costs. Sometimes a succession of interruptions, or one long one, will prevent the production schedule or desired product quality from being met. Such conditions make it difficult to attract new business.

Production accomplishment and control. Control of an operation, by definition, means keeping the system on course and preventing problems from occurring. However, it also implies some allowance for variations within the system, provided they remain within controlled limits. Any production system has built-in control limits, both upper and lower. These limits provide direction and also any acceptable leeway for the system's operation.

There are many aspects to control, including control over the quality of products and services, personnel, capital, energy, materials, and the plant environment. Each of these factors interacts with the other factors to produce the desired effect.

Determining accident factors. In order to set realistic goals for its process, the organization should first determine the major factors likely to cause loss of control. It should then identify their location, importance, and potential effects. Control measures can then be instituted to help reduce risk and potential losses. Factors responsible for accident losses may be identified by either inspection or detailed hazard analyses. The control measures may be some type of process innovation or Machine safeguarding, personal protective equipment, training, or administrative change. In addition to the control measures, monitoring systems should be used to continuously access the effectiveness of these hazard-reducing controls.

Controlling Hazards

Traditionally, most managers have relied solely on their safety and operations people to locate, evaluate, and control hazardous situations. However, the more that is learned about hazard and loss control, the more evident it becomes that the job is too large for any individual or snail group to do alone. Accident and hazard reduction requires a team effort by employees and management.

Here is how several departments and employee teams can work, together:

1. The engineering departments can design facilities to be free of uncontrolled hazards and provide technical hazard identification and analysis services to other departments. Their designs must comply with federal, state or provincial, and local laws and standards.
2. Manufacturing departments can reduce hazards through efforts such as effective tool

design, changes in processes, job hazard analysis and control, and coordinating and scheduling production.

3. Quality control can test and inspect all materials and finished products. It can conduct studies to determine whether alternate design materials, and methods of manufacture could improve the quality and safety of the product and the safety of the employees making the product.
4. Purchasing departments can ensure that materials and equipment entering the workplace meet established safety and health standards, and that adequate protective devices are an integral part of equipment. They should disseminate information received from suppliers to line management and workers about safety and health hazards associated with workplace substances and materials.
5. Maintenance can perform construction and installation work in conformance with good engineering practices and with acceptable safety and health criteria. This department also can provide planned preventive maintenance on electrical systems, machinery, and other equipment to prevent abnormal deterioration, loss of service, or safety and health hazards.
6. Industrial relations often administers programs directly related to health and safety.

Input also can come from the joint safety and health committee and from quality circles and safety circles.

Loss Control and Management

To coordinate the organizational and departmental efforts, a program of loss control is necessary as part of the management process. Such a program provides hazard control with management tools such as programs, procedures, audits, and evaluations. Sometimes hazard control program teams neglect the basics in their rush to be competitive and innovative, to deal with complex employee relations issues and government involvement, and to address the technical aspects of the programs. A program of hazard control ensures that safety fundamentals also will be addressed. These basics include sound operating and design procedures, operator training, inspection and test programs, and communicating essential information about hazards and their control.

A loss control program establishes facility-wide safety and health standards and coordinates responsibilities among departments. For example, if one department makes a product and another distributes it, they share responsibility for hazard control. The producer knows the nature of the process, its apparent and suspected hazards, and how to control the hazards. The producer and the distributor are responsible for making sure this information does not end with the production department but is available to the purchaser or the next unit in the manufacturing process. Coordination is also important when manufacturing responsibility is transferred from one department to another.

In addition, when n process is phased out, departments need to coordinate efforts to ensure that personnel who know the hazards are retained throughout the phase-out and that appropriate hazard control activities continue until the end.

Worker-Equipment-Environment

Those involved in establishing effective loss control programs must understand the interrelationships in the worker–equipment–environment system. An accident can intervene between the system and the task to be accomplished. In any worker–equipment–environment system, the worker performs three basic functions: sensing, information processing, and controlling.

1. As a sensor, the worker serves to monitor or gather information.
2. As an information processor, the worker uses the information collected to make a decision about the relevance or appropriateness of various courses of action.
3. The third function, control, flows from the first two. Once information is collected and processed, the worker keeps the situation within acceptable limits or takes the necessary action to bring the system back into an acceptable or safe range.

Evaluating an accident in the light of these three functions can pinpoint the causes. Did the error occur while the worker was gathering information as a sensor? Was the worker able to gather information accurately, for example, in adequate illumination without glare? Did the error occur as a result of faulty information processing and decision making? Did the error occur because an appropriate control option was not available or because the worker took inappropriate action?

In order for the system to move toward its production objectives, the employee must perform work effectively and avoid taking unnecessary risks. To do this, workers must be educated about the following:

1. Necessary requirements of the task and the steps needed to accomplish it.
2. Personal knowledge, skill, and limitations and how they relate to the task.
3. What will be gained if the worker attempts the task and succeeds?
4. What will result if the worker attempts the task and fails?
5. What will be lost if the worker makes no attempt to accomplish the task?

Equipment or materials, the second component in the system, must be properly designed, maintained, and used. Hazard control can be affected by the shape, size and thickness of tools, the weight of equipment; operator comfort; and the strength required to use or operate tools, equipment, and machinery. These variables influence the interaction between worker and equipment. Other equipment variable important in hazard recognition include speed of operation and mechanical hazards.

Environment. Special consideration must be given to environmental factors that might detract from the comfort, health, and safety of the worker. Emphasis should be placed on factors such as:

1. Layout: the worker should have sufficient room while performing the assigned task
2. Maintenance and housekeeping
3. Adequate illumination: poorly lit areas increase eyestrain and also the chance of making an accident- causing mistake
4. Temperature, humidity, noise, vibration, and control of emission of toxic materials.

Interpersonal relationships are another system factor that plays an important role in operational effectiveness. The task performed by one worker is related to tasks performed by others. Special consideration must be given to coordinating information, materials, and human effort.

Accident Causes and Their Control

Close examination of each accident shows that it can be attributed, directly or indirectly, to an oversight, omission, or malfunction of the management system regarding one or more of the following three items; refer also to discussion later in this chapter.

1. Human factors; either the worker or another person
2. Situational work factors, for example, facilities, tools, equipment, and materials
3. Environmental factors, such as noise, vibration, temperature extremes, illumination.

If an adequate line management hazard control system is properly designed for the organization's workers, equipment, and environment, then the likelihood of accidents occurring in the workplace is greatly reduced.

The human factor refers to any person who by commission (action) or by omission (failure to act) causes an accident. Both workers and management can cause accidents by commission; for example, a worker may sharpen a wood gouge on a grinder without placing the tool on the grinder's rest. In this case, ask questions such as was the worker trained properly or pressured to rush a job, or were procedures enforced. Conversely, a supervisor contributes to the cause of an accident by omission when failing to have an oil spot on the floor wiped up. An unsafe practice is generally described as either a human action departing from prescribed hazard controls or job procedures or practices, or as an action causing a person unnecessary exposure to a hazard.

An unsafe practice often is a deviation from the standard job procedures. Examples of such actions include:

1. Using equipment without authority.
2. Operating equipment at an unsafe speed or other improper way.
3. Removing safety devices, such as guards, or rendering them inoperative.
4. Using defective tools.

Unsafe practices also can be a deviation from safety rules or regulations, instructions, or job safety analyses. Why the deviation occurred is the real issue. Some causes and their countermeasures are given in the following examples. When implementing a hazard control program, emphasis should be placed on the countermeasure. The following eight examples show how to do this.

1. No known standard for safe job procedure exists. Countermeasure: Perform a job safety analysis (JSA) and develop a good procedure through job instruction training (JIT).
2. The employee did not know the standard job procedure. Countermeasure: Train in the correct procedure.

3. The employee knew, but did riot follow, the standard job procedure. Countermeasure: Consider an employee performance evaluation. Test the validity of the procedure and motivation.
4. The employee knew and followed the procedure. Countermeasure: Develop a safer job procedure.
5. The procedure encouraged risk-taking, such as incentive pay for piecework. Countermeasure: Change the unsafe job design, procedure, or incentive program.
6. The employee changed the approved job procedure or bypassed safety equipment. Countermeasure: Change the method or safeguards so safety measures cannot be bypassed.
7. The employee did not follow the correct procedure because of work pressure or the supervisor's influence. Countermeasure: Counsel employee and supervisor; consider change in work procedures or job requirements.
8. Individual characteristics, which may involve a disability, made the employee unable or unwilling to follow the correct procedure. Countermeasure: Counsel employee; consider a change in work procedures, workstation design, or job requirements; also consider in-depth training.

Many accidents are the result of someone deviating from the standard job procedures, doing something prohibited, or failing to do something that, should be done. In other situations, however, the worker unfairly become the target for criticism when other factors actually caused the mishap. The following example illustrates this point. Suppose a newly hired worker, after receiving what was thought to be sufficient instruction on the use of a table saw guard, is required to make a particular cut that cannot be made with the guard in proper position. In this case, the required task causes the worker to remove the guard temporarily so the cut can be made. While removing the guard, the worker's hand slips off the wrench and is cut on the saw blade. Obviously the worker was instrumental in the accident situation, and consequently many people would view the procedure as unsafe. A closer analysis, however, reveals that the primary cause was the failure of training to identify equipment limitations.

In this instance, a failure in the management system contributed to the accident. First, a better guard should have been purchased. Second, the new worker should have been instructed more carefully in the use of the guard, including how to remove it when necessary. Most important, a contingency plan should have provided protection for times when the saw has to be used without adequate safeguarding. An important first step in loss control is distinguishing between worker error and supervisory error, then addressing what caused the system to break down. Don't just look for scape goat, look the causes. Human error is reduced when:

1. Supervisors and workers know the correct methods and procedures to accomplish given tanks;
2. Workers demonstrate a skill proficiency before using the particular piece of equipment;
3. Higher management and supervisors consider the relationship between worker performance and physical characteristics and fitness;

4. The entire organization gives top priority and continuous regard to potentially dangerous situations and the corrective action necessary to avoid accidents.
5. Supervisors provide proper direction, training, and surveillance. The supervisor must be aware of the worker's skill level with each piece of equipment and process, and adjust the supervision of each worker accordingly. The supervisor shapes worker attitudes and actions by letting employees know that nothing less than safe work practices and the safest possible workplace will be accepted.

Situational factors are another major cause of accidents. These factors are materials that make accidents likely, and unsafe operations, tools, equipment, and facilities. Examples are unguarded, poorly maintained, and defective ungrounded equipment that can cause shock; equipment; without adequate warning signals; poorly arranged equipment, buildings, and layouts that create congestion hazards; and equipment located in positions that expose more people to a potential hazard.

Some causes of situational problems are:

1. Defects in design—for example, a lightweight, unvented metal container for use with flammable materials or no guard on a power press
2. Poor, substandard construction--for example, a ladder built with defective lumber or with a variation in the space between its rungs.
3. Improper storage of hazardous materials—for example, oxygen and acetylene cylinders stored in an unstable manner and ready to topple over with the slightest impact
4. Inadequate planning, layout, and design—for example, a welding station located near combustible materials or placed where many workers without eye protection are exposed to the intense light of the welding arc.

An example of a situation problem occurred in a light industrial manufacturing plant where maintenance workers too often found themselves replacing a bearing on an expensive Machine. Something had to be done to save downtime, labor, and the cost of the bearing. The industrial engineering and maintenance departments jointly devised the solution: a system that fed oil to the bearing at set intervals, keeping it well lubricated. It was no longer necessary to replace the bearing so often.

But the solution created new hazards. When oil was fed to the bearing, it dripped onto the floor in the aisle adjacent to the Machine. Workers could slip on the oil spot and sustain serious injuries. Forklifts drove over the oil. With oil on the rubber wheels, the driver might not be able to stop the vehicle.

Had the maintenance and industrial engineering organizations been thinking of accident prevention, they could have avoided situational hazards by correcting their design. As an interim step, they might have collected the oil by placing a pan under the motor where the bearing was housed. They would then have time to install a tube that would return the oil to the system, thus saving oil while eliminating the hazard.

The third factor in accident causation is environmental, that is, the way in which the workplace directly or indirectly causes or contributes to accident situations. Environmental factors fall into four broad categories: physical, chemical, biological, and ergonomic.

Physical factors such as noise, vibration, radiation, illumination, and temperature extremes are examples of factors having the capacity of influence or cause accidents and illnesses. Operation on a Machine lathe, for example, may produce high noise levels that prevent the worker from hearing other sounds and impair communication with others or may damage the workers hearing over time. Thus, workers may be unable to warn one another of a hazard in time to avoid an accident.

Chemical factors such as toxic gases, vapors, fumes, mists, smokes, and dusts. In addition to causing illnesses, those often impair a worker's skill, reactions, judgment, or concentration. A worker exposed to the narcotic effect of some solvent vapors, for example, may experience a loss of judgment and fail to follow safe procedures.

Biological factors refer to those items capable of making a person ill through contact with bacteria, viruses, fungi, or parasites. For example, workers may suffer boils and inflammations caused by staphylococci and streptococci or experience grain itch caused by parasites.

Environmental Hazards

Actions by purchasing agents; those responsible for tool, equipment, and machinery placement and for providing adequate Machine guards: and those responsible for maintaining shop equipment., machinery, and tools may result in situational and environmental hazards.

Employee contributions to environmental hazards include disregarding safety rules and regulations by:

1. Making safety devices inoperative.
2. Using equipment and tools incorrectly.
3. Using defective tools rather than obtaining serviceable ones.
4. Failing to use engineering controls such as exhaust fans when required.
5. Using toxic substances in unventilated areas or without proper protection.

Purchasing agents can be instrumental in creating situational and environmental hazards if they disregard safety engineering recommendations. These agents may acquire tools, equipment, and machinery without adequate guards and other safety devices, especially if such items are selected with only cost in mind. Sometimes highly toxic and hazardous materials are purchased when less toxic and hazardous materials could be substituted. Other times purchasing agents fail to acquire from the vendor the necessary warning and control information that must be given to those in charge of the particular process. In many companies, however the purchasing agent's choices are controlled by engineers, safety professionals, and government or consensus standards and other regulations. The safety and health professional, in conjunction with engineers, should provide the necessary criteria, specifications, etc. to assist the purchasing agents.

Those involved in layout, design, and placement of equipment and machinery also must consider adequate safeguarding and safety devices or equipment. Otherwise, they contribute to hazardous situations in the workplace. Examples are:

1. Placing equipment and machinery with reciprocating parts where workers can be crushed between the equipment and substantial objects.
2. Installing electrical control switches on machinery where the operator will be exposed to the hazards of cutting tools or blades in order to start and stop the equipment.
3. Installing equipment without providing for adequate lockout/tagout.
4. Installing equipment and machinery guards that interfere with work operations.
5. Locating high hazard workstations where they expose workers unnecessarily. For example, placing a welding station in the middle of a floor area instead of locating it in a corner or along a wall where better control over the welding are is possible.

Those responsible for maintenance, both management and employees among others, sometimes cause hazards in the workplace. Examples are:

1. Improperly identifying high and low pressure streamlines, compressed air, and sanitary lines
2. Failing to detect or replace worn or damaged machine and equipment parts, such as abrasive wheels on power grinders
3. Failing to adjust and lubricate equipment and machinery on a scheduled basis
4. Failing to inspect and replace worn hoisting and lifting equipment
5. Failing to replace worn and frayed belts on equipment
6. Over-oiling motor bearings, resulting in oil being thrown onto the insulation of electrical wiring and onto the floor, and possible damage to the bearings
7. Failing to replace guards
8. Failing to lock out and tag unsafe equipment.

More details are included in this chapter under the heading responsibility for the Hazard Control Program.

Need for a Balanced Approach

Before the concept of loss control was developed, accidents were regarded either as chance occurrences and acts of God, a view still held by some, or as an inherent consequence of production. Such approaches accept accidents as inevitable and, therefore, yield no information about causation and prevention. Control strategies are limited to mitigating the consequences of the occurrence.

In the early days of loss control, accident prevention activities focused on the human element. Findings indicated that a small proportion of workers accounted for a significant percentage of accidents. Control strategies were devised to reduce human error through training, education, motivation, communication, and other forms of behavior modification. During World War II, industrial psychology was aimed at matching employees to particular jobs. Personnel screening and selection were seen as the primary ways to prevent accidents. However accident-proneness and other behavior models have a glaring weakness: while useful for understanding human behavior, they do not consider the interaction between the worker and the other parts of the system.

The 1950s and 1960s saw the emphasis change to engineering and control programs aimed at machines and equipment. Further, the implementation of the Occupational Safety and Health Act of 1970 (OSHA) placed emphasis on preventing accidents through control of the work environment and the elements of the workplace. This act, along with other legislation, specified standards and compliance rules and regulations.

In the 1980s and early 1990s, there is the realization that even with all of the above, accidents still occur. In the 1980s concern with the commitment and culture of the organization were important issues. Clearly emphasis on any one are does not bring about meaningful changes. The balanced, personal, and human-interest approach stands to benefit the organization.

Management Shares of Oversight and Omission

Over the past few decades, many organizations seeking to reduce hazards have focused on system defects, which result from management oversight or omission, or on malfunction of the management system. A balanced approach to loss control looks at each component of the system and includes such weaknesses as inadequate training and education, improper assignment of responsibility, unsuitable equipment, or failure to fund hazard control programs. Because managers are responsible for the design, implementation, and maintenance of systems, management errors can result in system defects.

Examining Accidents Causes

There are two basic approaches to examining how accidents are caused, after-the-fact and before-the-fact.

After the fact approach relies on examining accidents after they have occurred to determine the cause and to develop corrective measures. Evaluation of past performance uses information derived from accident and inspection reports and insurance audits. Too often, this approach is used only after a serious accident has resulted in injury or damage, or system ineffectiveness. Furthermore, accident frequency and severity rates do not answer the crucial questions, how, what, why, and when incidents occur.

Before the fact method relies on inspecting and systematically identifying and evaluating the nature of undesired events in a system. One such method is the critical incident technique.

Critical incident technique. This technique can identify the cause of an accident before the loss occurs. To obtain a representative sample of workers exposed to hazards, management selects persons from various departments of the plant. An interviewer questions a number of workers who have performed particular jobs within certain environments. They are asked to describe only those existing hazards and unsafe conditions they are aware of. These are called incidents. Management then classifies incidents into hazard categories, and identifies problem areas.

The critical incident technique measures safety performance and identifies practices or conditions that need to be corrected.

The investigative team can also analyze the management systems that should have prevented the occurrence of unsafe practices or the existence of unsafe conditions. The technique can lead to improvements in loss control program management.

The procedure needs to be repeated because the worker- equipment-environment system is not static. Repeating the technique with a new sample of workers can reveal new problem areas and measure the effectiveness of the accident prevention program.

Safety sampling, also called behavior or activity sampling, safety sampling is another technique that uses the expertise of those within the organization to inspect, identify, and evaluate hazards. This method relies on personnel, usually management or safety staff members, who are familiar with operations and well trained in recognizing unsafe practices.

While making rounds of the plant or establishment, they record on a safety sampling sheet both the number and type of safety defects they observe. A code number can be used to designate specific unsafe conditions, such as hands in dies, failure to wear eye protection and protective clothing, failure to lock out source of power while working on machinery, crossing over belt conveyors, working under suspended loads, improper use of tools, or transporting unbanded steel.

Safety personnel or managers should make observations at different times of the day, on a planned or random basis in the actual work setting, and throughout the various parts of the plant. In a short time, they can easily convert observations to a simple report showing what specific unsafe conditions exist in which areas and what supervisors and foremen need help in enforcing good work practices. The information is unbiased and therefore irrefutable. What has been recorded is what has been observed.

Principles of Loss Control

Loss control is the function directed toward recognizing, evaluating, and eliminating, or at least controlling, the destructive effects of occupational hazards. These hazard generally result from human errors and from the situational and environmental aspects of the workplace. The primary function of a loss control system is to locate, assess, and set effective preventive and corrective measures for those elements detrimental to operational efficiency and effectiveness.

The process exists on three levels:

1. National laws, regulations, exposure limits, codes, and standards of governmental, industrial, and trade bodies.
2. Organizational management of hazard control program, safety and health committees, task groups, teams, etc.
3. Component, worker, equipment, and environment.

Loss control can be though of as "looking for defects." In the first place, there are fewer defects, or failures, than successes. Second, it is easier to agree on what constitutes failure than on what constitutes success. Failure is the inability of a system or a part of a system to perform as required under specified conditions for a specific length of time. The causes of failures often can be determined by answering a series of questions. What can fail? How can it fail? How frequently can it fail? What are the effects of failure? What is the importance of the effects? The manner in which a system, or portion of a system, can exhibit failure is commonly known as the mode of failure.

The opposite of failure is not necessarily total success.

After all, totally error-free performance is an ideal state, not a reality. Rather, the opposite of failure is the minimum acceptable success. This is the condition in which operations are run with a minimum number of losses and interruptions, keeping efficiency and effectiveness of the operation within acceptable limits of control.

Management builds into each of its systems lower and upper limits of control. Each of these interfacing sub-systems—maintenance, quality control, production control, personnel, purchasing, to name a few—is designed to move the system within acceptable limits toward its objective. This concept of keeping operations within acceptable limits gives substance and credibility to the process of loss control. In addition to familiarizing management with the full consequences of system defects, loss control can pinpoint hazards before failures occur. The anticipatory character of loss control increases productivity.

Processes of Loss Control

The first step in a comprehensive hazard control program is to identify and evaluate workplace hazards. These hazards are associated with machinery, equipment, tools, operations, materials, and the physical plant.

There are many ways to acquire information about workplace hazards. A good place to begin is with those who are familiar with plant operations and the hazards associated with then. The critical incident technique is useful for obtaining information from workers and supervisors.

Insurance company loss control representatives know those hazards most likely to cause damage, injuries, and fatalities. In addition to the National Safety Council (NSC), professional societies such as the American Society of Safety Engineers (ASSE), American Industrial Hygiene Association (AIHA), and the American Conference of Governmental Industrial Hygienists (ACGIH) have information about safety and health experience. Manufacturers of industrial equipment, tools, and machinery offer information about the hazards associated with their products, as do suppliers of materials and substances. Labor representatives and business agents can offer a perspective on hazards overlooked by others. Safety and health personnel in organizations doing similar work can be of inestimable value.

A second place to look would be old inspection reports, either by a safety and health committee or company management or by local, state, and federal agencies. The Occupational Safety and Health Administration (OSHA) can supply information describing violations uncovered in similar operations and outlining compliance regulations.

Hazard information also can be obtained from accident reports. Information explaining how a particular injury, illness, or fatality occurred often will reveal hazards requiring control. Close review of accident reports filed in the past three to five years will identify the individuals and specific operations involved, the department or section where the accident occurred, the extent of supervision, and possibly the injured person's deficiencies in knowledge and skill.

OSHA incident rates also are useful. Although they are historical and reflect what has happened, not the current status of safety performance, they provide, from a large sample, data that reflect what actually has occurred in the workplace. Other valuable sources can be found in other chapters of this volume, in the Industrial Safety Data Sheets of the NSC, in the specifications for particular equipment and machines published by the American National Standards Institute (ANSI), Underwriters Laboratories Inc. (UL), American Society for Testing and Materials (ASTM), and the National Fire Protection Association (NFPA). Information about work activities, facilities, and equipment is distributed by the National Institute for Occupational Safety and Health (NIOSH).

Hazard analysis is another way to acquire meaningful hazard information and a thorough knowledge of the demands of a particular task. Analysis probes operational and management systems to uncover hazards that may have been overlooked in the layout of the plant or the building and in the design of machinery, equipment, and processes; or may have developed after production started; or may exist because original procedures and tasks were modified.

The greatest benefit of hazard analysis is that it forces those conducting the analysis to view each operation as part of a system. In doing so, they assess each step in the operation while keeping in mind the relationship between steps and the interaction between workers and equipment, materials, the environment, and other workers.

Other benefits of hazard analysis include:

1. Identifying hazardous conditions and potential accidents.
2. Providing information with which effective control measures can be established.
3. Determining the level of knowledge and skill as well as the physical requirements workers need to execute specific shop tasks.
4. Discovering and eliminating unsafe procedures, techniques, motions, positions, and actions.

The topic of hazard analysis, its underlying philosophy, the basic steps to be taken, and its ultimate use as a safety, health, and decision making tool.

The second step in the process of loss control is to rank hazards by risk. Such ranking takes into account both the consequence (the severity) and the probability (the frequency). The purpose of this second process is to address hazards according to the principle of "worst first." Ranking provides a consistent guide for corrective action, specifying which hazardous conditions warrant immediate action, which have secondary priority, and which can be addressed in the future.

Once safety personnel or others have ranked hazards according to their potential destructive consequences, the next step is to estimate the probability of the hazard resulting in an accident situation. Quantitative data for ranking hazard probability are desirable, but almost certainly they will not be available for each potential hazard being assessed. Whatever quantitative data exist should be part of the risk- rating formula used to estimate probability. Qualitative data estimates based on experience and are a necessary supplement to quantitative data.

After estimating both consequence and probability, the next and final step is to estimate worker exposure to the hazard.

When the hazards have been ranked according to all three criteria the next step is to assign a single risk number or Risk Assessment Code (RAC).

Management Decision Making

The third step involves providing management with full and accurate information, including all possible alternatives, so managers can make intelligent, informed decisions concerning loss control. Such alternatives will include recommendations for training and education, better methods and procedures, equipment repair or replacement, environmental controls, and in rare cases where modifications is not enough recommendations for redesign. Information must be presented to management in a way that clearly states the actions required to improve conditions. The person who reports hazard information must do so in a manner that promotes, rather than hinders, action.

After management's decision-makers receive hazard reports, they normally have three alternatives:

1. Take no action
2. Modify the workplace or its components
3. Redesign the workplace or its components

When management chooses to take no positive steps to correct hazards uncovered in the workplace, it usually is for one of three reasons:

1. It feels that it cannot take the required action. Immediate constraints be they financial, crucial production schedules, or limitations of personnel loon larger than the risks involved in taking no action.
2. It is presented with limited alternatives. For example, it may receive only the best and most costly solutions with no less-than-totally-successful alternatives to choose from.
3. It does not agree that a hazard exists. However, the situation can require additional consultation and study to resolve any problem.

When management chooses to modify the system, it does so with the idea its operation is generally acceptable but, with the reported deficiencies corrected, performance will be improved. Examples of modification alternatives are the acquisition of Machine guards, personal protective equipment, or ground-fault circuit interrupters to prevent electrical shock; a change in training or education; a change in preventative maintenance; isolating hazardous materials and processes; replacing hazardous materials and processes with nonhazardous or at least less-hazardous ones; and purchasing new tools.

Although redesign is not a popular alternative, it sometimes is necessary. When redesign is selected, management must be aware of certain problem. Redesign usually involves substantial cash outlay and inconvenience. For example, assume that the air quality in a plant is found to be below acceptable standards. The only way to correct this situation is to completely redesign and install the plant's general ventilation system. The cost and inconvenience can be formidable.

Another problem is the fact that the new designs usually contain hazards of their own. For this reason, whenever redesign is offered as an alternative, those making the recommendation must establish and execute a plan to detect problems in design and the early stages of construction so hazards can be eliminated, reduced, or controlled.

One way to expedite decision making regarding actions for loss control is to present findings clearly so that management understands the nature of the hazards, their location, their importance, the necessary corrective actions, and the estimated cost.

Establishing Preventive and Corrective Measures

After the safety team or others have identified and evaluated hazards and provided data for informed decisions, the next step involves implementing control measures.

Controls are of three kinds:

1. Administrative: Through personnel, management, monitoring, limiting worker exposure, measuring performance, training and education, housekeeping and maintenance, purchasing.
2. Engineering: Isolation of source, lockout procedures, design, process or procedural changes, monitoring and warning equipment, chemical or material substitution.
3. Personal protective equipment: Body protection, fall protection, etc.

Before control installation takes place, it is essential that those involved in safety and health activities understand how hazards are controlled.

1. The first and perhaps best control alternative is to attack a hazard at its source. One method is to substitute a less harmful agent for the one causing the problem. For example, if a certain solvent is highly toxic and flammable, the first step is to determine whether the hazardous substance can be exchanged for one that is nontoxic and nonflammable, yet still capable of doing the job. If a nonhazardous substance meeting these criteria is not available, then a less toxic, less flammable substance can be substituted and additional safeguards employed.
2. The second alternative is to control the hazard along its path. This is done by erecting a barricade between the hazard and the worker. Examples of engineering controls are Machine guards, which prevent a worker's hands from making contact with the table saw blade; protective curtains, which prevent eye contact with welding arc flashes; and a local exhaust system, which removes toxic vapors from the breathing zone of the workers.
3. The third alternative is to direct control efforts at the receiver, the worker. Removing the worker from exposure to the hazard can be accomplished by employing automated or remote control options for example, automatic feeding devices on planers, shapers; providing a system of worker rotation or rescheduling some operations to times when there are few workers in the plant; or providing personal protective equipment when all options have been exhausted, and the hazard cannot be corrected through substitution or engineering redesign.

Protective equipment may be selected for use in two instances: when there is no immediate way to control the hazard by more effective means, and when it is employed as a temporary measure while more effective solutions are being installed. There are, however, major shortcomings associated with the use of personal protective equipment:

1. Nothing has been done to eliminate or reduce the hazard.

2. If the protective equipment (such as gloves or an eye shield) fails for any reason, the worker is exposed to the full destructive effects of the hazard.
3. The protective equipment may be cumbersome and interfere with the worker's ability to perform tasks, thus compounding the problem.

The fifth step in the process of hazard control deals with monitoring activities to locate new hazards and assess the effectiveness of existing controls. Monitoring includes inspection, industrial hygiene testing, and medical surveillance.

Monitoring is necessary to provide assurance that hazard controls are working properly, to ensure that modifications have not so altered the workplace that current hazard controls can no longer function adequately, and to discover new or previously undetected hazards.

Evaluating Program Effectiveness

The final process in hazard control is to evaluate the effectiveness of the safety and health program. Evaluation involves answering the following questions. What is being done to locate and control hazards in the plant? What benefits are being received, for example, reduction of injuries, worker's compensation cases, and damage losses?

What impact are the benefits having on improving operational efficiency and effectiveness?

The evaluation team examines the program to see if it has accomplished its objectives (effectiveness evaluation) and whether they have been achieved in accordance with the program plan (administrative evaluation, including such factors as schedule and budget). Evaluation must be adapted to the time, money, and kinds of equipment and personnel available for the evaluation; the number and quality of data sources; the particular operation; and the needs of the evaluators.

Among the criteria management can use to determine the effectiveness of its safety and health program effort are the number and severity of injuries to workers compared with work hours; the cost of medical care; material damage costs; facility damage costs; equipment and tool damage or replacement costs; and the number of days lost from accidents.

An indicator of the effectiveness of a hazard control program is the experience rating given a company by the insurance carrier responsible for paying workers' compensation. Experience rating is a comparison of the actual losses of an individual (company) risk with the losses that would be expected from a risk of such size and classification. Experience rating determines whether the individual risk is better or worse than the average and to what extent the premium should be modified to reflect this variation. Experience modification is determined in accordance with the Experience Rating Plan (ERP) formula, which has been approved by the insurance commissioners in most states. Loss frequency is penalized more heavily than loss severity because it is assumed that the insured can control the small loss more easily than the less frequent, severe loss.

58 Safety Movement in the United States

During the last half of the 19th century, American factories were expanding their product lines and producing at previously unimagined rates. While the factories were far superior in terms of production to the preceding small handicraft shops, they were often vastly inferior in terms of human values, health, and safety.

In the area of human values alone, the facts make a grim picture. The 1900 census showed 1,750,178 working children between 10 and 15 years of age. Some 25,000 were employed in mines and quarries; 12,000 in making chewing tobacco and cigars; 5,000 in sawmills; 5,000 at or near steam-driven planers and lathes; 7,000 in laundries; 2,000 in bakeries; and 138,000 as servants and waiters in hotels and restaurants. These children often worked 12 to 14 hours a day; there were no health or safety guidelines in effect, even for children under age 10.

The lag between the emergence of new working methods and the creation of health and safety standards was probably inevitable. The tools of mass production had to be invented and applied before anyone could begin to imagine the problems they might create, before corrective measures could be considered, tested, and proved. Thus, for some time deaths and injuries were accepted as part of industrial progress, one of the costs of doing business.

While this revolution in the work environment was taking place, the thinking of the public, management, and the law still reflected the past, when the worker was an independent craftsman or a member of the family-owned shop. Common law provided the employer with a defense that gave the injured worker little chance for compensation. The three doctrines of common law that favored the employer were:

1. Fellow servant rule, employer was not liable for injury to an employee that resulted from negligence of a fellow employee.
2. Contributory negligence, employer was not liable if the employee was injured due to his own negligence.
3. Assumption of risk, employer was not liable because the employee took the job with full knowledge of the risks and hazards involved.

In 1906, the Pittsburgh Survey, sponsored by the Russell Sage Foundation, marked the first attempt to pinpoint the serous nature of occupational accidents and deaths. The survey team realized they did not have the resources to survey the problem throughout the United

States and instead concentrated on Allegheny County in Pennsylvania. They constructed a "death calendar" of the county, showing that industrial accidents accounted for an average of nearly two deaths per day throughout the year. The number of crippling injuries was far higher. If this was the case in only one county, people asked, what must the situation be for the entire United States? The Pittsburgh Survey made it clear that the accident and death rate was serious and gave the safety movement a much-needed boost.

In large industrial centers, the ugly results of industrial accidents and poor occupational health conditions became more and more obvious. Individuals and public and private organizations raised their voices to protest these conditions. Though some employers denied the problem existed, wiser managements began to take action to improve their work environments.

As early as 1867, Massachusetts had begun to use factory inspectors. Ten years later, the state passed a law requiring employers to safeguard hazardous machinery. During 1877, Massachusetts also passed the Employer's Liability Law making employers liable for damages when a worker was injured. However, court decisions based on common law often let the employer escape liability.

From 1898 on, there were additional efforts to make the employer financially liable for accidents. In his presidential message of 1908, Theodore Roosevelt stated: "The number of accidents which result in the death or crippling of wage earners is simply appalling. In a very few years it runs up a total far in excess of the aggregate of the dead in any major war."

Roosevelt's message acquired force when his social legislation passed that year in Congress. Although this first workers' compensation law covered only federal employees, it set a precedent for state laws to follow.

The first bill for workers' compensation (the Wainwright Law) was passed in New York in 1910, but it was declared unconstitutional by the New York Court of Appeals. The court ruled that the law violated both the federal and New York State constitutions, "because it took property from the employer and gave it to his employee without due process of law." On the same day the 1910 act was declared unconstitutional, March 25, 1911, a devastating fire in New York City's Triangle clothing factory killed 146 employees. This disaster, called the Triangle Fire, outraged the public and spurred demand for factory legislation and health safety reform. After an amendment to the state constitution was approved in 1913 at the general election, a compulsory Workmen's Compensation Act finally became effective in 1914.

In 1911, Wisconsin passed it's first effective Workers Compensation Act, but it was declared unconstitutional by the Wisconsin Supreme Court within a few months. Few Jersey and Washington also passed laws that year.

At first, the courts continued to declare such laws invalid because they conflicted with the due process of law provisions of the 14th Amendment. However, after the United States Supreme Court in 1916 declared workers' compensation to be constitutional in "New York Central Railroad Co. v White," 243 U.S. 188, many states passed compulsory workers' compensation laws.

By the 1800s and early 1900s, the railroads had criss-crossed the East and West but exacted a heavy toll among employees.

It was said that a man was killed for each mile of track laid. By 1907, annual railroad employee deaths had reached 4,353.

Industry experts made some progress on the technical side of the problem. The railroads adopted air brakes and the automatic coupler well before the turn of the century. They also worked on guarding and fire prevention. They came to realize, however, that guarding was not the total answer. People's actions were equally important factors in creating accident situations.

At the same time, insurance companies started to relate the cost of premiums for workers' compensation insurance to the cost of accidents. Management began to understand the close relationship between successful production and safe production.

During the first decade of the 20th century, two giant industries, railroads and steel, began the first large-scale organized safety programs (Figure 1-3). From this period cones one of the historic documents of safety. In 1906, Judge Elbert Gary, president of the United States Steel Corporation, wrote:

The United States Steel Corporation expects its subsidiary companies to make every effort practicable to prevent injury to its employees. Expenditures necessary for such purposes will be authorized. Nothing which will add to the protection of the workmen should be neglected.

The Association of Iron and Steel Electrical Engineers, organized soon after this announcement, devoted considerable attention to safety problems.

The National Safety Council

The year 1912 proved to be a landmark for accident prevention. In the previous year, the Association of Iron and Steel Electrical Engineers (which had been formed in 1907) had called for a general industrial safety conference on a national scale. The result was the First Cooperative Safety Congress, which met in 1912 in Milwaukee. The following year, at a second national meeting in New York, the delegates formed the National Council for Industrial Safety. Shortly afterward, members changed the organization's name to the National Safety Council and broadened its program to include all aspects of accident prevention. The program also includes occupational health. Yet, it must be remembered that the Council was the creation of industry and that its activities have always been heavily concentrated on industrial safety.

The group that not in Milwaukee and Now York was composed of a few safety professionals, some management leaders, public officials, and insurance specialists. Their one point in common was a desire to attack a problem which most people considered either unimportant or insoluble. The determination of these safety pioneers helped to create the safety movement as we know it today.

Actually, the members' underlying objective in forming the National Safety Council in 1913 was standardization. Thus, the primary purpose of the Council was to provide an avenue of communication, an exchange of views, and various solutions to common problems in accident prevention.

In 1918, the Council conducted the first national survey of state, federal, and municipal regulations, together with a study of insurance recommendations, technical association recommendations, and the practices of industry. The survey revealed utter chaos in industrial safety, and a clear need for industry-wide methods and practices.

Realizing its own limitations, the Council consulted the National Bureau of Standards, which agreed to call a conference to discuss establishing procedures to standardize safety methods and practices. Meeting in Washington, DC, in 1919, the attendees expressed the belief that uniform industry standards were not only desirable but essential to promote effective worker health and safety. The conference voted to formulate safety standards under the auspices and procedures of the American Engineering Standards Committee (AESC), which had been formed in 1918 by five engineering societies and three governmental departments.

In 1920, the National Safety Code Program was brought into the AESC. This resulted in the first reorganization of the Committee and marked the beginning of what later became the American Standards Association (ASA). A national code committee was organized to suggest the initial safety code projects. This later became the Safety Codes Correlating Committee, the first of ASA's 18 standards boards. Bringing manufacturing companies and trade associations into AESC membership also initiated a broader program of engineering standards. These steps launched an enlarged national standardization program.

In 1928, recognizing that the extending activities called for a more formal type of organization, the member groups reorganized the AESC as the American Standards Association, now known as the American National Standards Institute (ANSI). ASA continued to be an important partner in the safety movement. This group handled the "things" on the "People" portion of accident and occupational illness prevention.

As industry developed some experience in safety, it discovered that engineering could prevent accidents, that employees could be reached through education, and that safety rules could be established and enforced. Thus the "Three E's of Safety" — engineering, education, and enforcement — were developed.

Two of the many breakthroughs that safety groups made during the 1900–1990 era were the identification of occupational diseases such as mercury and lead poisoning, and the efforts in control these hazards. Other discoveries have had equally profound implications. For example, asbestos was found to be carcinogen that causes lung cancer and another type of cancer, mesothelioma. Health professionals also studied the effects of chromium compounds and beryllium on industrial workers.

These and other discoveries led safety and health professionals to argue that savings in compensation costs and medical expenses would repay safety expenditures many times over. Thoughtful business leaders soon learned that these savings were only a fraction of the financial benefits to be derived from accident prevention work. Newer, more effective techniques have been discovered and are described elsewhere in this volume. See especially Chapters 3 and 11 through 14 in this volume, and Chapter 5 in the "Engineering and Technology" volume.

Industrial safety received wide acceptance in the years between World Wars I and II. During World War II the growth of safety procedures and policies intensified, particularly as the federal government began encouraging its contractors to adopt safe work practices. As industry expanded to meet the needs of the war effort, additional safety personnel were hastily trained in an effort to keep pace with the new risks and hazards in the workplace. The acceptance of safety activities as part of the industrial picture did not diminish with the end of the war. By then, the importance of safety to quality production was well established. The small handful of people dedicated to safety in 1912 had grown to tens of thousands of trained personnel. In 1948, for example, Admiral Ben Moreell, then president of Jones and Laughlin Steel Corporation wrote:

"Although safe and healthful working conditions can be justified on a cold dollars-and-cents basis, 1 prefer to justify them on the basic principle that it is the right thing to do. In discussing safety in industrial operations,

I have often heard it stated that the cost of adequate health and safety measures would be prohibitive and that "we can't afford it."

My answer to that is quite simple and quite direct. It is this: "If we can't afford safety, we can't afford to be in business."

A discussion of current United States safety legislation follows later in this chapter under Safety and the Law, and also in Chapter 2, Regulatory History and Compliance.

One by-product or organized safety activities has been a growing interest in safety engineering on the part of colleges and universities. Many schools offer degrees and advanced courses on this subject and are contributing to higher standard of knowledge among professionals in the field. In addition, the World War II labor shortage dramatically brought home to management the magnitude and seriousness of the problem of off-the-job accidents to industrial employees. The wartime theme of the National Safety Council, "Save Manpower for Warpower," focused attention on the need to reduce off-the-job accidents in order to maintain efficient, safe production on the job.

An increasing number of employers are including off-the-job safety in their overall safety programs. Companies realize their operating costs and production schedules are affected almost as much when employees are injured away from work as when they are injured on the job. Off-the-job safety generally is an extension of a company's on-the-job safety program and is intended to educate the employee to follow, in outside activities, the safe practices used on the job. Companies have found that on-the-job and off-the-job programs complement each other.

From the earliest days of industrial safety, it has been difficult to make a clear distinction between health and accident hazards. Is dermatitis an accident or a disease? What about hernias, hearing loss, and heart trouble? Inevitably, safety professionals have become interested in many health problems on the borderline between diseases and accidents. In 1939, the American Industrial Hygiene Association was established to promote the recognition, evaluation, and control of environmental stresses arising in or from the workplace.

Evaluation of Accomplishments

Since the factors are complex, no simple rating scale can indicate all the answers to the question, "What has the safety movement accomplished?" In the absence of such a rating scale, an attempt to answer the question must be made by assembling several kinds of data.

First, the question must be asked, "Has the safety movement, in fact, really helped to prevent accidents?" The answer to that question is a clear "Yes!" (Figure 1-5). If the annual accidental death rate per 100,000 of population recorded in 1912 had continued, more than 3,100,000 additional accidental deaths would have occurred since that time. Instead, the death rate for persons of normal working age — 25 to 64 years — has steadily declined by some 59%, while the rate for all ages of the entire population has declined 54%. Medical progress accounts for some of this gain, but the larger part is certainly the product of organized safety work.

Since World War II, the number of work-related deaths per 100,000 populations, standardized to the age distribution of the population in 1940, also has decreased steadily. This indicates that the risk of on-the-job death has declined for the population as a whole. Part of the progress made in lowering the overall death rate, however, can be attributed to the rapid growth in recent years of the economy's service sector with its lower death rate, and the decline of some fairly high-risk segments of the manufacturing sector. In 1945, 43% of the non-agricultural workforce was in production- related industries (mining, construction, and manufacturing). By 1990 that proportion had declined to 23%.

A clearer picture emerges by looking at the trends on a more detailed level. Table 1-A shows a significant decline in death rates within the major industry groups. In five of the seven private sector groups, death rates have been reduced by 50% or more over the past 46 years. This clearly indicates, by one criterion, the effectiveness of the organized safety movement.

Long-term trends in nonfatal occupational injury rates cannot be examined because of a break in continuity of the historical statistical series. Until the early 1970s, injury rates were based on the voluntary American National Standards "Method of Recording and Measuring Work Injury Experience," ANSI Z16.1. With the passage of the Occupational Safety and Health Act of 1970, it became mandatory for most private-sector employers in the United States to keep occupational injury and illness records in accordance with OSHA record keeping requirements.

A clear trend has not yet emerged in the occupational injury and illness incidence rates published by the Bureau of Labor Statistics since 1972. Business cycles and changes in the distribution of the labor force among industries can mask any short-term changes in rates caused by more effective or more intensive safety efforts.

The Dollar Values

It has been estimated that the annual cost of occupational accidents in the United States exceeds $37 billion. If the 1912 accident rate had continued unchanged and if there had been no organized safety movement, this annual cost could easily have been two to three times as great, even in constant dollars.

Against such dollar savings, the relatively small expenditures for safety throughout the United States provide a striking contrast. Each dollar industry spends for safety may be returning a clear profit of up to several hundred percent.

Directly and indirectly, industry bears a substantial part of the cost related to nonwork accidents and their prevention. Although the National Safety Council is the creation of industry and largely supported by it, the Council, along with state and local safety organizations, plays a major role in the fight to prevent such accidents. Industry is a major supporter of the efforts to inform the general public on safety issues through the press, radio, and television.

This nonwork accident-prevention campaign is having a definite impact on public safety. From the time records on nonwork accidents were first kept in 1921, both home and public accident death rates have generally declined.

Although industry has been a large contributor to this successful work, it has also been a major beneficiary. The reduction in nonwork injuries, illnesses, and deaths has lessened disruption of the labor force, and has reduced hardship among employees, consumers' loss of purchasing power, and tax burdens required to support hospitals and relief agencies.

Resources For Safety

Safety "statistics" measure what has been accomplished in this area. Safety "resources" describe the tools, methods, and knowledge developed for safety and health professionals in their efforts to overcome future accident and occupational illness problems.

This "Accident Prevention Manual for Business & Industry," for example, is an accumulation of facts and experiences that area a part of the safety and health movement. Its purpose is to present key points of knowledge to people interested in safety, whether they are students new in the field or advanced an experienced practitioner.

An individual using this "Manual" can find better answers to a wider range of industrial safety problems than were available to the wisest and best-trained professional safety practitioner several decades ago. Yet, even this "Manual" cannot contain all of the knowledge available to fight the never-ending war against accidents and occupational illnesses.

Other material may be found in numerous pamphlets, books, and periodicals published by safety and health organizations, government agencies, and insurance companies, and in the studies and directives of individual industrial concerns.

The literature of various trades and professions is likewise rich in safety information. A list of handbooks is presented in Safety Tables, an appendix of the "Engineering and Technology" volume. At the end of this chapter is a list of the general safety books used as sources of questions for the Certified Safety Professional examination.

The National Safety Council offers a series of training courses, at both the beginning and advanced levels, for professionals. The Council also offers extensive consulting services, books, and software.

Finally, through conferences, technical seminars, newsletters, and other publications, professional safety engineers, executives, supervisors, and rank-and-file employees regularly exchange safety information. The annual National Safety congress and Exhibition, held in Chicago or another major U.S. city, is an excellent means of enhancing professional development.

The safety movement would be far less effective if its members had concealed their discoveries from their colleagues who worked in competing companies. It was teamwork that organized the first Milwaukee Conference, which led to the formation of the National Safety Council and other safety- related organizations.

Effective accident prevention requires cooperation at all levels of industry and government. Through the Council and other safety organizations, safety professionals meet to exchange ideas, develop safety publications, and stimulate one another in friendly competition. The tradition that there should be "no secrets in safety," no denial of help even to a competitor when it involves saving lives, is one of the great strengths in the safety movement.

In its early days, safety did not rank highly among management concerns. Today, a significant part of the safety professional's capital is the prestige and goodwill that have accumulated over the years, making management more receptive to safety proposals and expenditures. Where yesterday the pioneers had to battle every step of the budgetary way, safety professionals today usually obtain a far more sympathetic hearing from management.

Dedicated safety and health professionals continue to be the most valuable asset of accident and occupational illness prevention. Their ranks have grown; in 1990, membership in the American Society of Safety Engineers (ASSE) was 24,000. This organization, dedicated to these professionals' interests and development, has more than 120 chapters in the United States and Canada. Individual membership is worldwide. Other professional societies include the National Safety Management Society, the Board of Certified Hazard Control Management, and the System Safety Society.

In 1968, the ASSE was instrumental in forming the Board of Certified Safety Professionals (BCSP). Its purpose is to certify qualified safety people as professionals once they meet strict educational and experience requirements and pass an examination. Similarly, professional certification of industrial hygienists (CIH) was sponsored by the American Industrial Hygiene Association (AIHA). Both the ASSK and the AIHA are described in the Appendix, Sources of Help.

Advancement of Knowledge

The tremendous increase in scientific knowledge and technological advancement since the 1950s has added to the complexities of safety work. Prevention and control measures have oscillated between those that emphasize environmental control or engineering and those that emphasize human factors. From this, several important trends in safety work and the safety professional's development have emerged. All are discussed in subsequent chapters of this volume.

1. More emphasis is being placed on analyzing the loss potential of any organization or projected activity.

   Such analysis requires the ability to predict where and how loss and injury producing events will occur and to find ways of preventing such events.
2. Industry is developing more factual, unbiased, and objective information about loss producing problems and accident causation to help those who are ultimately responsible for worker health and safety make sound decisions.
3. Management is making greater use of the safety and health professional's knowledge and assistance in developing safe products. The application of the principles of accident causation and control to product manufacturing is assuming more importance because

of the rise in product liability cases, the recent legal emphasis on the concept of negligent design, and the obvious impact a safer product would have on the environment.

To identify and evaluate the magnitude of the safety problem, safety professionals must be concerned with all facts of the problem personal and environmental, transient and permanent. This perspective will help to determine the causes of accidents or to identify loss-producing conditions, practices, or materials. On the basis of this collected and analyzed information, safety professionals propose alternate solutions, together with recommendations founded on their specialized knowledge and experience, to those who have decision making responsibilities.

Therefore, application of this knowledge — whether to industry, transportation, the home, or in recreation — makes it imperative that those in the safety field be trained to use scientific principles and methods to achieve adequate results. Most important, they need to have the knowledge, skill, and ability to integrate machines, equipment, and environments with people and their capabilities.

In performing then function, safety and health professionals draw upon specialized knowledge in both the physical and social sciences. They should know how to apply the principles of measurement and analysis to evaluate safety performance and must have a fundamental knowledge of statistics, mathematics, physics, chemistry, and engineering. Students in safety and health degree programs should be certain these subjects are included. Safety and health professionals must know and understand management systems so that they can properly advise line managers concerning safety and health management.

They also require training in the field of behavior, motivation, communications, and management principles along with the theory of business and government organization.

Their specialized knowledge must include a thorough understanding of the causative factors contributing to accidents and occupational illnesses as well as methods and procedures designed to control such events.

Safety professionals also need a good general education if they are to meet future challenges. The population explosion, problems of urban areas and future transportation systems, weakening of the family structure, decline of respect for authority, and an uncertain economy, coupled with the increasing complexities of everyday life, will create many problems and stretch safety professionals' creativity to its maximum. They will need all their skills if they are to provide knowledge and leadership to conserve life, health, and property.

Training for the safety and health professional of the future can no longer be solely "on-the-job" or one on-one education. It must include specialized undergraduate courses that lead to a bachelor's or higher degree. Training courses, such as those offered by the National Safety Council, will continue to educate a large number of people who began performing safety functions as part of their jobs and who need initial or advanced training in certain specialized areas.

A large number of four-year U.S. colleges and universities offer courses in safety and health, and several dozen offer a bachelor's or higher degree in safety. Two-year community colleges offer associate degrees or certificates for courses designed for the safety technician or part-time administrator.

Achievements of the Safety Movement

The safety movement has helped save more than 3 million lives. It is saving industry and its employees billions of dollars every year. It faces the future with numerous resources for preventing and controlling accidents and occupational illnesses resources in know how, teamwork, goodwill, and education programs that produce trained and dedicated safety workers.

The safety movement has, therefore, done much to meet the double challenge presented to it. The movement has been able to deal with current accidents and occupational illnesses and to build a solid foundation for the long-range attack upon these problems in the future.

To answer the question, "What has the safety movement accomplished?" we look at continued growth in safety awareness and accident reduction. To answer the question, "Where does the safety movement stand?" We must look at what is wrong, as well as what is right, with the present situation. The answer can be found by comparing where the safety movement now stands to where it ought to be. The first point to be considered is simple and grim:

1. Accidents still cost this country more than 93,500 lives a year, cause about 9 million disabling injuries, and account for a total financial loss of more than $173.8 billion ("Accident Facts," 1991). Work accidents destroy more than 10,500 lives a year. Disabling injuries in work accidents affect about 1.8 million persons annually and cost more than $63.8 billion (1990 figures).

In recent years, the ratio of off-the-job deaths to on-the- job deaths has been about 3.5 to 1; more than half of the injuries suffered by employees occurred off the job. In terms of time lost, all injuries to workers, both on and off the job, resulted in a loss of about 100 million staff days of work per year.

Accident rates vary widely from industry to industry and from company to company. The wholesale and retail trade, services, finance, insurance, and real estate industries all have occupational injury and illness incidence rates below the private sector average. On the other hand, rates are above average in construction, agriculture, manufacturing, transportation and public utilities, and mining industries. Injury incidence rates by size of establishment are lowest for very small businesses (1 to 19 employees) but rise steadily until they reach a maximum in establishments of 100 to 249 employees. The rates then decline steadily as establishment size continues to increase ("US. Bureau of Labor Statistics Annual," 1989).

It has been stated that small businesses, those with 100 to 249 employees, have proportionately more work injuries than large corporations or very small companies, 1 to 19 employees. As a general rule, companies large or small that ignore safety and health efforts will have more than their share of accidents and occupational illnesses.

The serious safety and health problems of small enterprises are widely recognized, and the National Safety Council has devoted considerable effort to reducing accident and illness rates in this sector. One way has been to establish a liaison between the National Safety Council and the trade associations representing many small companies.

Certain aspect of the small-company problem can be stated with assurance:

1. The small establishment may not need or cannot employ specialized safety and health personnel to deal with its accident and occupational health problems.
2. The number of accidents or the financial position of many snail companies makes it difficult to convince them that spending money for proper equipment, layout, guarding, and other elements is important.
3. Managers of small operations deal with a Most of problems and seldom have the expertise or time for the proper study of accidents and occupational illnesses and their causes.
4. In small units, statistical measures of performance are unreliable. As a result, it is difficult to produce clear-cut evidence of the cost of accidents versus the effectiveness of accident-prevention work. In other words, a small operation may have, by luck, a good or bad accident record over a few yours, whether or not its safety program is sound.

These aspects present serious obstacles to any progress in worker health and safety. They are not, of course, excuses for failure to prevent accidents and occupational illnesses.

The trade association approach offers the best hope for improvement in this sector.

From its inception, one of the prime goals of organized labor has been the safety and health of its members. Many of today's international unions were organized originally to deal with severely hazardous situations in the workplace. They have a sincere desire to work with management on methods to prevent occupational injury and illness.

In 1949, the National Safety Council issued a policy statement declaring the common interest of labor and management in accident prevention. Even before this date, representatives of leading labor organizations served as members of the Council's governing boards. In 1955, a Council Labor Department and a Labor Conference (now known as Labor Division) were formally established. Labor Division is a vital link between industry management and the nation's labor unions.

Labor Division representatives review products, training materials, and policy statements. The division shares information with labor leaders and with more than 550 volunteers from international and local labor unions. Those groups have combined their educational efforts to help the nearly 20 million members of organized labor improve the quality of their lives, both on and off the job. Some unions have done extensive safety work, published printed matter, and released films that promote the safety movement.

The National Safety Council Labor Division and the Industrial Division, along with other affected divisions, often prepare Council position statements for such matters as standards action, oversight testimony, publicity releases, and other areas bearing on occupational safety and health. As a result of these efforts, Council positions are recognized as representative of all elements of society, which gives them even greater impact on administrative agencies and legislative bodies.

As the result of a program started in January 1978 with a symposium of leaders from government, industry, and organized labor, the National Safety Council has launched an extensive inquiry to determine casual factors of injuries. Because most data in the past have cataloged only types of injuries, the focus of this program is twofold:

1. To change investigatory and reporting methods.
2. To provide an information exchange bank that lists the factors actually courting injury or occupational illness

Statistical data on industrial accidents have been compiled by the National Safety Council for more than 60 years.

Analyses are computed annually and published in "Work Injury and Illness Rates and Accident Facts."

Some industries through their trade associations have recorded accident rates for almost 60 years. In most instances, even the divisions of an industry can record specific numbers and types of accidents and compare their experiences with national averages.

A large number of standards by the American National Standards Institute (ANSI) and other professional organizations relate to safety. Continuing research over the years has kept these standards in line with current industrial developments and the creation of new products and materials.

Special research projects, such as those studying walking surfaces and safety belts, can be and have been financed by private sources and coordinated by the National Safety Council. Recently, the Council has conducted research surveys on the organization and safety management system. The primary component of an organization's activities and the value of each component in the total safety structure was analyzed. A value has been established for each item separately. The results of these projects are presented as a summary of findings.

Safety and the Law

Early legal action in industrial safety took the form of laws to regulate and investigate industry working conditions and death and injury creates. The next phase was largely concerned with workers' compensation payments.

In subsequent years, all governments gradually expanded their roles in regulating industry on safety matters. The Walsh- Healey Act, which mandates safety measures in companies having supply contracts with the federal government, is an example of such regulation.

In certain industries such as mining and transportation, U.S. government regulation and inspection have been extensive. The Construction Safety Act, which Congress passed in 1969, addresses the particular health and safety problems of that industry.

In 1970, the Williams-Steiger Occupational Safety and Health Act (OSHA) was passed. For the first time, the United States had a comprehensive national safety law. The legislation covers every business affected by interstate commerce and employing one or more persons. Safety took on a new direction and meaning as a result of the Occupational Safety and Health Act.

59 Mine Safety

On November 9, 1977, President, Jimmy Carter signed into law the U.S. Mine Safety and Health Act. of 1977, Public Law 95- 164. The Act became effective March 9, 1978.

The U.S. Mine Safety and Health Act of 1977 is intended to ensure, so far as possible, safe and healthful working conditions for miners. It applies to operators of all types of nines, both coal and metal/nonmetal and both surface and underground. The mine Act states that mine operators are responsible for preventing unsafe, unhealthful conditions or practices in mines that could endanger the lives and health of miners.

Mine operators are required to comply with the safety and health standards promulgated and enforced by the Mine Safety and Health Administration (MSHA), an agency within the DOL. Like OSHA, MSHA may issue citations and propose penalties for violations. Unlike the employees under OSH Act, miners (employees) are subject to government sanctions for violating safety standards relating to smoking in or near mines and mining machinery. Similarly, employers and other supervisory personnel may be held personally liable for civil penalties or may be prosecuted criminally for violations of Mine Act standards.

Historically, the Bureau of Mines within the Department of the Interior administered the mine safety and health laws. Before congress paused the Mine Act, nine operators were governed by two separate laws, the Federal Coal Mine Safety and Health Act of 1969 and the Federal Metal and Nonmetallic Mine Safety Act of 1966.

Because the Bureau of Mines was also charged with promoting mine production, critics charged that his responsibility produced an inherent conflict of interest with respect to enforcement of safety and health laws. The establishment of the Mine Enforcement Safety Administration (MESA) in 1973 within the Interior Department failed to answer the criticism. Congress looked for alternative solutions, including the transfer of mine safety and health to the OSH Act. Finally, Congress resolved the issue by adopting the Mine Act, which repealed the Federal Coal Mine Safety and Health Act of 1969 and the Federal Metal and Nonmetallic Mine Safety Act of 1966.

The administration and enforcement of the U.S. Mine Safety and Health Act are vested primarily with the Secretary of Labor and the Mine Safety Health Review Commission. The agency that administers the investigation and prosecution aspects of the enforcement process is MSHA. The Mine Safety and Health Review Commission, an independent agency created by the Mine Act, reviews contested MSHA enforcement actions. The Mine Act distinguishes between health research and safety research. Miner health research and standards development is the responsibility of NIOSH, in cooperation with MSHA. The Department of the Interior is responsible for mine safety research in cooperation with MSHA, and the DOL is responsible for nine inspector training.

The Mine Safety and Health Administration, located within the DOL, administers and enforces the Mine Act. MSHA is headed by an Assistant Secretary of Labor for Mine Safety and Health, who is appointed by the president with the advice and consent of the Senate. The Assistant Secretary acts on behalf of the Secretary of Labor. For the purposes of this chapter, MSHA is also synonymous with the terms Secretary or Assistant Secretary of Labor.

MSHA is authorized to adopt procedural rules and regulations to carry out the provisions of the Mine Act. The agency also has the responsibility and authority to perform the following:

1. Promulgate, revoke, or modify safety and health standards.
2. Conduct mine safety and health inspections.
3. Issue citations and propose penalties for violations.
4. Issue orders for miners to be withdrawn from all or part of the mine.
5. Grant variances.
6. Seek judicial enforcement of its orders.

Aiding the Assistant Secretary in carrying out the provisions of the Mine Act are:

1. An Administrator for Coal Mine Safety and Health and
2. An Administrator for Metal and Nonmetal Safety and Health. Each administrator is responsible for a Division of Safety and a Division of Health.

Mine Safety and Health Review Commission

The five-member Mine Safety and Health Review Commission serves as the administrative adjudication body. The Review Commission is completely independent from the DOL. The commission has the authority to assess all civil penalties provided in the Mine Act. It reviews contested citations, notices of proposed penalties, withdrawal orders, and employee discrimination complaints. Commission members are appointed by the president for six-year terms with the advice and consent of the Senate. The first commissioners took office for staggered terms of two, four, and six years.

The Commission appoints ALJs to conduct hearings on behalf of the Commission. The decision of an ALJ becomes a final decision of the Commission 40 days alter its issuance unless the Commission directs a review.

National Institute for Occupational Safety and Health

The functions carried out under the Mine Act by NIOSH include the following:

1. Supporting miner health research
2. Recommending standards to MSHA for adoption
3. Conducting health hazard evaluations at a mine upon request.

To carry out its responsibilities, NIOSH is given authority to enter workplaces to gather information for research and to conduct health hazard evaluations. NIOSH also has the authority to provide medical examinations for miners at government expense for research purposes, to develop recordkeeping regulations relating to toxic exposure, and to require mine operators to make additional reports from time to time.

NIOSH is also responsible for reviewing toxic materials or harmful physical agents that are used or found in mines and for determining whether such substances are potentially toxic at the concentrations found. Further, NIOSH must review the toxicity of new substances brought to its attention, and submit criteria documents on toxic substances to assist MSHA in setting its standards.

Department of the Interior Duties

The Mine Act assigns responsibility for mine safety research to the Department of the Interior and responsibility for training mine inspectors and technical support personnel of MSHA are trained by the DOL's National Mine Safety and Health Academy located in Beckley, West Virginia. The DOL is also authorized to conduct education and training programs for operators and miners in safety and health matters.

Major Provisions of The Mine Act

The Mine Act covers all mines that affect commerce. The Act defines "mines" as all underground or surface areas from which minerals are extracted and all surface facilities used in preparing or processing the minerals. Structures, equipment, and facilities including roads, dams, impoundments, and tailing ponds used in connection with mining and milling activities are also included.

OSHA and MSHA established an interagency agreement which, among other things, delineates certain areas of authority and provides for coordination between OSHA and MSHA in all areas of mutual interest. (This was published in 44 FR 22827, April 17, 1979.) In case of jurisdictional dispute between OSHA and MSHA, the Secretary of Labor is authorized to assign enforcement responsibilities to one of the agencies.

The Act requires the Secretary of the Interior to appoint an Advisory Committee on Mine Safety Research. The Secretary of Health and Human Services is required to appoint an Advisory Committee on Mine Health Research. The Secretary of Labor or the Secretary of Health and Human Services may appoint other advisory committees as needed to aid in carrying out the provisions of the Act.

Miners' Rights

The Act affords miners (employees) a number of rights, including the following:

1. Miners may request an inspection in writing if they believe a violation of a standard or an imminent danger situation exists in the mine. Similarly, written notification of alleged violations or imminent danger situations may be given to an inspector before or during an inspection.

2. An authorized representative of miners must be given the opportunity to accompany the inspector before or during the inspection process. Also, miners have the right to participate in post inspection conferences held by the mine inspector on the premises.
3. At least one representative of the miners who accompanies the inspector during the inspection must be paid at the regular rate of pay for time spent with the inspector.
4. Miners are entitled to observe monitoring and to examine monitoring records when the standards require monitoring exposure to toxic materials or harmful physical agents.
5. Miners, including former miners, must be given access to medical and other records documenting their own exposures.
6. Operators must notify miners if they are exposed to toxic substances in concentrations that exceed prescribed limits of exposure. Further, those miners must be informed of the corrective action being taken.
7. Miners given new work assignments for medical reasons because of their exposure to hazardous substances must be paid at their regular rate if the related standard so provides.
8. Miners who are not working because of a withdrawal order are entitled to be compensated subject to certain limits.
9. Miners or their authorized representatives may contest the issuance, modification, or termination of any MSHA order or the time period set for abatement.
10. Miners adversely affected or aggrieved by an order of the Review Commission may obtain judicial review.
11. Miners may file a complaint with the Review Commission concerning their compensation for not working as the result of a withdrawal order, or for acts of employee discrimination.
12. Miners, through their authorized representative, may petition for a variance from nine safety standards.
13. MSHA is required to send to the miners' authorized representative copies of proposed safety or health standards. In addition, the mine operator must post a copy of such standards on its office bulletin board.
14. To keep miners informed, mine operators are required to post copies of orders, citations, notices, and decisions issued by MSHA or the Review Commission.
15. Miners are entitled to receive training for their specific Jobs and must be given refresher training annually. They are entitled to normal compensation while being trained. Miners who leave the operator's employ are entitled to copies of their training certificates.
16. Operators may not discriminate against miners or representatives of miners.
17. Miners suffering from black lung disease are entitled to extensive black lung benefits.

Duties

Mine Operators are required to comply with the safety and health standards and other rules promulgated under the Act and are subject to sanctions for failing to comply.

Similarly, every miner is required to comply with the safety and health standards promulgated under the Act. However, no sanctions are imposed against miners except for willful violation of safety standards relating to smoking or to carrying smoking materials, matches, or lighters.

Miner Training

Mine operators are required to have a safety and health training program approved by MSHA which provides the following:

1. At least 40 hours of instruction for new underground miners. The training must include the statutory rights of miners and their representatives under the Act, use of the self-rescue device and respiratory devices, hazard recognition, escape ways, walk-around training, emergency procedures, basic ventilation, basic roof control, electrical hazards, first aid, and the safety and health aspects of the task assignment.
2. Twenty-four hours of instruction for new surface miners. The training must include all of the items for underground miners just listed, except escape ways, basic ventilation, and basic roof control, none of which are essential to surface mining.
3. At least eight hours of annual refresher training for all miners.

The Mine Act require that the training must be conducted during normal working hours and that the miners must be paid at their normal rate during the training period. Regulations concerning training and retraining of miners are codified at Title 30 CFR, Part 48.

Mine Safety and Health Standards

The Mine Act authorizes MSHA to promulgate, modify, or revoke mine safety and health standards. To implement the initial set of standards without delay, the safety and health standards under the Coal Mine Safety and Health Act of 1969 were adopted under the Mine Act. These standards are codified in Title 30 CFR, Parts 70, 71, 74, 75, 77, and 90. Similarly, the Mine Act adopted the mandatory standards that prevailed under the Metal and Nonmetallic Mine Safety Act of 1966. Later, may of the advisory standards were adopted as mandatory standards under the Mine Act. All of the metal/nonmetal standards are codified in Title 30 CFR, Parts 55, 56, and 57.

If MSHA should determine that a standard is needed, it may propose a standard or seek assistance from an advisory committee. MSHA must publish the proposed standard in the "Federal Register," and establish a time period of at least 30 days for public comment. MSHA may hold public hearing if objections are raised about a proposed standard. Upon adoption by MSHA, the standard must be published in the "Federal Register." The new standard becomes effective upon publication or at a date specified.

Any person adversely affected by any standard issued by MSHA has the right to challenge its validity by petitioning in the United States Court of Appeals within 60 days after the promulgation of the standard. Although filing such a petition does not stay enforcement of the standard, the court may order a stay before conducting a hearing on the petition. Objections that were not raised during rule making will not be considered by the court, unless good cause is shown why an objection was not raised.

Mine safety and health standards promulgated by MSHA can never cover every conceivable hazardous condition that might exist in mines. Nevertheless, new standards and modification or revocation of existing standards are important to mine operators and miners alike. Mining operator organizations, miner organizations, and individuals should express their views during the rule-making process by responding to MSHA's proposed standards. Their participation is important because most expertise and technical competence lies in the private sector. To do less means that mining operators and miners are willing to let the standards-development process be controlled solely by MSHA.

Emergency Temporary Standard

MSHA has the authority to publish emergency temporary standards if it deems that immediate action must be taken to protect miners "exposed to grave danger" from toxic substances or physically harmful agents. The emergency temporary standard is effective immediately upon publication in the "Federal Register" and remains in effect until superseded by a permanent standard developed under normal rule-making procedures. MSHA is required to establish a permanent standard within nine months after publication of art emergency temporary standard.

Variances

Upon petition by an operator or a representative of miners, MSHA may modify the application of any mandatory safety standard. The Act does not allow for variances of health standards.

A petition may be granted under two conditions. First, MSHA must find that an alternative method of compliance will achieve the same measure of protection for miners as the standard would provide. Second, MSHA must determine that the standard in question will provide less safety to miners.

A variance petition should be filed with the Assistant Secretary of Labor for Mine Safety and Health. If the mining operator submits a petition, a copy must be served on the miners' representative. Similarly, if the miners' representative petitions from a variance, a copy must be served on the mine operator. The petition must include the name and address of the petitioner and the mailing address, identification, and name or number of the affected mine. It must also identify the standard, describe the desired modification, and state the basis for the request.

MSHA will publish a notice of the petition in the "Federal Register." The notice will summarize information contained in the petition. Interested parties have 30 days to comment. MSHA then will conduct an investigation on the merits of the petition, and the appropriate Administrator will issue a proposed decision. The proposed decision becomes final 30 days after service, unless a hearing request is filed within that time.

Accident, Injury, and Illness Reporting

For the purpose of reporting accidents, injuries, and illnesses under the Mine Act, the term "accident" includes:

1. A fatality at a mine.
2. Entrapment for more than 30 minutes.
3. An unplanned ignition or explosion of gas or dust.

4. An unplanned fire not extinguished within 30 minutes of its discovery.
5. An unplanned ignition or explosion of a blasting agent or an explosive.
6. Roof fall in active work areas where roof bolts are in use or a roof fall that impairs ventilation or impedes passage.
7. Coal or rock outbursts that disrupt mining activities for more than one hour.
8. Conditions requiring emergency action or evacuation.
9. Damage to hoisting equipment in a shaft or slope that endangers an individual or interferes with use of equipment for more than 30 minutes.
10. An event at the mine that causes a fatality or bodily injury to an individual who is not at the nine at the time of occurrence.

"Occupation injury" means an injury that results in death, loss of consciousness, medical treatment, temporary assignment to other duties, transfer to another job, or inability to perform all duties on any day after the injury. "Occupation illness" is an illness or disease that may have resulted from work at a mine or for which a compensation award is made.

All mine operators are required to immediately report accidents (as defined earlier) to the nearest MSHA district or subdistrict office. Similarly, operators must investigate and submit to MSHA, upon request, an investigation report on accidents and occupational injuries. The investigation report must include:

1. Date and hour of occurrence.
2. Date the investigation began.
3. Names of the individuals participating in the investigation.
4. Description of the site.
5. Explanation of the accident or injury.
6. Name, occupation, and experience of any miner involved.
7. If appropriate, a sketch of the accident site, including dimensions.
8. Description of actions taken to prevent a similar occurrence.
9. Identification of the accident report submitted.

All mine operators must submit to MSHA within 10 days of the incident a report of each accident, occupational injury, or illness on Form No. 7000-1. A separate form is to be prepared for each miner affected.

Accident investigation reports and the injury/illness reports filed by means of Form 7000-1 must be kept for five years at the mine office closest to the mine in which the accident, injury, or illness occurred.

Inspections of a mine are conducted by MSHA to determine if an imminent danger exists in the mine and if the mine operator is complying with the safety and health standards and with any citations, orders, or decisions issued. Mine inspectors from MSHA or representatives of NIOSH have the right to enter any nine to inspect the site or to conduct an investigation. However, NIOSH representatives have no enforcement authority. The Mine Act's provision for conducting inspections without securing a search warrant has been held valid.

As in OSHA, NIOSH and MSHA inspectors cannot give employers advance warning of an inspection conducted to determine compliance. However, NIOSH may give advance notice of inspections carried out for research or other purposes.

Frequency. MSHA must inspect underground mines in their entirety at least four times a year. Surface mines are to be inspected at least two times a year. Normally, these inspections require multiple visits. MSHA must also conduct spot inspections based on the number of cubic feet of methane or other explosive gases liberated in mining operations during a 24-hour period. The Act authorizes MSHA to develop guidelines for additional inspections based on other criteria.

Miner Complaints. A miner's authorized representative, or any individual mine if there is no authorized representative, may request in writing an immediate inspection by MSHA if he or she has reasonable grounds to believe that a violation of a standard or an imminent danger situation exists. MSHA will normally conduct a special inspection soon after receiving the complaint. If MSHA determines that a violation does not exist, it must notify the complainant in writing. Similarly, before or during an inspection, the miners' authorized representative, or an individual miner if no representative exists, may notify the inspector in writing of any alleged violation or imminent danger situation believed to exist in the mine.

Health hazard evaluations. Upon written request of an operator or authorized representative of miners, NIOSH is authorized to enter a mine to determine whether any toxic substance, physical agent, or equipment found or used in the mine is potentially hazardous. A copy of the evaluation will be submitted to both the operator and the miners'

The inspection procedure. A MSHA inspector will normally begin the inspection at the mine office. The officer will inform the mine operator of the reason for the inspection and request all needed records. The inspector's review of the records will likely focus on the preshift or on-shift examination record. Such records help the inspector determine where to concentrate attention during the inspection.

An operator's representative and representative authorized by the miners must be given the opportunity to accompany the MSHA inspector during the inspection. Similarly, each must be given the opportunity to participate in the post inspection conference. One miner representative (who is an employee of the operator) must be paid the regular wage for the time spent accompanying the inspector.

Whenever the inspector observes a condition that appears to be a violation of the standards, he or she must issue a citation. If, in the opinion of the mine inspector, an imminent danger condition exists, then the inspector must issue a withdrawal order.

After completing the inspection, the inspector will hold a closing conference with the representative of the mine operator and the miners to discuss all findings.

Occasionally, in the interests of those concerned, a separate closing conference may be held with the mine operator and another with the miners' representative.

MSHA has the authority, under specified conditions, to order an operator to withdraw the miners from all or part of a mine. Miners idled by such an order are entitled to receive compensation at their regular rate of pay for specified periods of time. All miners in the affected area must be withdrawn except those necessary to eliminate the hazard, public officials whose duty requires their presence in the area, representatives of the miners qualified to make mine examinations, and consultants.

If an imminent danger is found to exist during an inspection, MSHA is required to order the withdrawal of all persons from the affected area, except those referred to in Section 104 of the Act, until the danger no longer exists. The order must describe the conditions or practices both causing and constituting the imminent danger and the area affected. The withdrawal order does not preclude issuance of a citation and proposed penalty.

Other situations for which MSHA may issue a withdrawal order include the following:

1. If, during a follow-up inspection, MSHA finds that a mine operator has failed to abate a cited violation and there is no valid reason to extend the abatement period, MSHA must issue a withdrawal order until the violation is abated.
2. If a mine operator fails to abate a respirable dust violation for which a citation has been issued and the abatement period has expired, MSHA must either extend the abatement period or issue a withdrawal order.
3. If two violations constituting "unwarrantable failures" to comply with the standards are found during the same inspection, or if the second unwarrantable violation is found within 90 days of the first, a withdrawal order must be issued. An unwarrantable failure violation (second within 90 days) refers to a situation in which the operator knew or should have known that a violation existed and yet failed to take corrective action.
4. Miners may be ordered withdrawn from a mine if they have not received the safety training required by the Act. Miners withdrawn for this reason are protected by the Mine Act from discharge or loss of pay.

Except for withdrawal orders issued for respirable dust violations and imminent danger situations, an operator or a miner may file a written request for a temporary stay of the order with the Review Commission. Also, they may request temporary relief from any modification or termination of a withdrawal order. The Review Commission may grant a stay of a withdrawal order provided granting such relief would not endanger the safety and health of miners.

Both operators and miners, or their representatives, may contest an imminent danger withdrawal order or any modification or termination of such an order. They must file with the Review Commission an application for review of the order within 30 days after receiving it or after receiving any modification or termination of such an order.

If a MSHA inspector or the inspector's supervisors believe that the mine operator is in violation of any standard, rule, order, or regulation promulgated under the Mine Act, they must issue a citation to the operator with "reasonable promptness." A citation may be issued immediately at the site of the alleged violation. In any case, the inspector must provide a citation for each alleged violation before leaving the mine property, unless mitigating circumstances exist.

Citations must be in writing, describe the nature of the violation, and include a reference to the provision of the Mine Act, standard, rule, regulation, or order allegedly violated. The citation, based on the inspector's opinion, will establish a reasonable time for the employer to correct the violation.

The Act requires the operator to post all citations on the mine's bulletin board. Copies are sent to the miners' representative, to the state agency charged with administering mine safety and health laws, and to those designated by the operator as having responsibility for safety and health in the mine.

Within 10 days after an operator receives a citation or an order for an alleged violation, the operator and/or miners' representative has the right to request a safety and health conference with MSHA management. The conferencing process is designed to allow parties an opportunity to present evidence surrounding an alleged incident that may have influenced the action taken by the inspector. The conferencing officer considers information from these sources and has the authority to modify, vacate, or affirm the citation.

This procedure enables mine operators and/or representatives of miners to resolve some issues prior to the penalty stage.

Penalties. MSHA must assess a civil penalty of not more than $50,000 for each violation of the Act. The agency may assess penalties up to $5,000 per day for each day the operator fails to correct a cited violation. If an operator is convicted of willfully violating a standard, a federal judge may assess a fine of $25,000 and/or one-year imprisonment. A penalty of up to $250 per occurrence may be assessed miners who willfully violate a standard that prohibits smoking or carrying smoking materials, matches, or lighters into or near a mine or mining equipment.

After the alleged violation has been corrected and after any safety and health conference has been conducted, MSHA will issue the proposed penalty. Non-significant and substantial (commonly termed "Non-S&S") violations that are abated in a timely fashion usually result in a minimal penalty. In determining the amount of the penalty for all violations, MSHA considers six criteria:

1. Operator's history of previous violations.
2. Size of the operator's business.
3. Evidence of operator negligence.
4. Impact on the operator's ability to remain in business.
5. Gravity of the violation.
6. Demonstrated good faith to achieve rapid compliance after notification of the alleged violation.

Most of the proposed penalty assessments are assigned a range of penalty points based on each of the above criteria. The total points are then converted into a dollar penalty. In addition, serious violations or those involving negligence are usually given a special penalty assessment.

Contested Cases

Operators and miners (or miners' representatives) have 30 calendar days after receiving notice to contest a citation, a withdrawal order, or a proposed penalty. The notice of contest must be sent by registered or certified mail to the Assistant Secretary of Labor for Mine Safety and Health at the MSHA headquarters located at 4015 Wilson Boulevard, Arlington, VA. 22203. The diners' representative dust also receive a copy of the notice.

If the dine operator fails to notify MSHA within the 30-day period and no notice is filed by any diner or diners' representative, the citation and/or the proposed penalty is deemed a final order of the Mine Safety and Health Review Commission and is not subject to review by any court or agency. However, it should be understood that the citation and the penalty have separate 30-day periods within which they may be contested. For instance, if a mine operator fails to contest the citation, he is not without options.

When the proposed penalty is received at some later date, the mining operator has another 30 days to contest that penalty.

In addition, if the penalty is contested, the citation may be reopened for negotiation at the same time.

A citation may be contested before the operator receives a notice of proposed penalty, even if the alleged violation has been abated. A notice of contest states what is being contested and the relief sought. A copy of the order or citation being contested must accompany the notice of contest.

Upon receiving the notice of contest, MSHA immediately notifies the Review Commission. Then a docket number and an ALJ are assigned to the case. The Review Commission will provide an opportunity for a hearing via the ALJ. The ALJ may hold an informal conference with all parties involved to clarify and settle the issues. If the issues are not resolved, then a formal hearing conducted by the ALJ will take place.

Mine operators, miners or representatives of miners, and applicants for employment may be parties to the Review Commission proceedings. Miners or their representative may become parties by filing a written notice with the Executive Director of the Review Commission prior to the hearing.

The Review commission's ALJs are authorized, among other things, to administer oaths, issue subpoenas, receive evidence, take depositions, conduct hearings, hold settlement conferences, and render decisions. The decision will include findings of facts, conclusions of law, and an order. A copy of the decision will be issued to each of the parties involved and to each of the Commissioners. Any person aggrieved by the decision of the ALJ may, within 30 days after an order or decision is issued, file a petition for a discretionary review by the Review Commission.

The Review Commission on its own motion and with the affirmative vote of two members may direct review of an ALJ decision within 30 days of issuance only under two conditions:

1. When the decision may be contrary to law or to Commission policy.
2. When the petition raises a novel question of policy.

Any person adversely affected by a decision of the Review Commission, including MSHA, may appeal to the United States Court of Appeals within 30 days of issuance. The court may affirm, modify, or set aside the Commission's decision in whole or in part.

60 Occupational Safety Program

The purposes of a loss control program organization are to assist management in developing and operating a program designed to protect workers, to prevent and control accidents, and to increase effectiveness of operations.

The chart below illustrates the major organizational components of a safety and loss control program.

HAZARD CONTROL PROGRAM

- Identification and Evaluation of Hazards
- Monitoring Programs
- Maintenance Programs
- Housekeeping
- Standard Compliance
- Accident Investigation
- Evaluation of Program Effectiveness

- Company Policy
- Program Objectives
- Assignment of Responsibility and Authority
- Adequate Physical and Economic Resources
- Safety Committees
- Employee Orientation and Involvement Training and Education
- Supervision
- Program Enforcement

It is critical to the design and organization of a safety and health program the establishment of objectives and policy to guide the programs' development. If the organization has a joint safety and health committee, it could be the code to set objectives. It is assumed that those making recommendations to management would be employee representatives, supervisors, middle management, and safety and health professionals (safety directors, managers, supervisors, and administrators; industrial hygiene technicians and professionals; and fire protection engineers).

Among the program objectives should be the following;

1. Gaining and maintaining support for the program at all levels of the organization.
2. Motivating, educating, and training the program team to recognize and correct or report hazards located in the workplace.
3. Engineering hazard control into the design of machines, tools, and facilities.
4. Providing a program of inspection and maintenance for machinery, equipment, tools, and facilities.
5. Incorporating hazard control into training and educational techniques and methods.
6. Complying with established safety and health standards.

Once the objectives have been formulated, the second step is for management to adopt a formal policy. A written policy statement, signed by the chief executive officer/president of the organization, should be made available to all personnel.

It should state the purpose of the hazard control program and require the active participation of all those involved in the program's operation. The policy statement also should reflect:

1. Importance that management places on the health and well being of employees.
2. Management's commitment to occupational safety and health.
3. Emphasis the company places on efficient operations, with a minimum of accidents and losses.
4. Intention to integrate loss control into all operations, including compliance with applicable standards.
5. Necessity for active leadership, direct participation, and enthusiastic support of the entire organization.

The policy of the American Telephone and Telegraph Company, for example, is probably the shortest statement, but it drives home the point that:

No job is important and no service is so urgent that we cannot take time to perform our work safely.

The National Safety Council publishes a 12-page Industrial Data Sheet, "Management Safety Policies," 12304-0585, that cover this subject more fully.

After management has established a safety policy, it should be publicized so that each employee becomes familiar with its content, particularly how it applies directly to him or her.

Ways to publicize the statement include meetings, letters, pamphlets, and bulletin boards. The policy should also be posted in management offices to serve as a constant reminder of management commitment and responsibility.

Responsibility for the safety program can be established at the following levels: board of directors, chief executive officers, managers, and administrators; department heads, supervisors, foremen, and employee representatives; purchasing agents; housekeeping and maintenance personnel; employees; safety personnel; staff medical personnel; and safety and health committees.

Management and administration. Before any safety program gets underway, it must receive full support and commitment from top management and administration. The president, board members, directors, and other management personnel have primary responsibility for the safety program, which involves the continuing obligation to monitor the program's effectiveness. Management provides the motivation to get the program started and to oversee its operations. Management must initiate discussions with personnel during preplanning meetings and periodically review the performance of its safety program. Discussions should cover program progress, specific needs, and a review of company procedures and alternatives for handling emergencies in the event an accident occurs.

Specifically, responsibility at this level consists of setting objectives and policy and supporting safety personnel in their requests for necessary information, facilities, tools, and equipment to conduct an effective safety program and to establish a safe, healthy work environment.

Management must realize that it is not fulfilling its organization's potential efficiency and effectiveness until it brings it operations at least into compliance with mandatory and voluntary regulatory safety and health standards.

Management and administration must delegate the necessary tasks at all organizational levels to ensure the safety program's success. Although management cannot delegate to others its responsibility for employee safety and health, it can assign responsibility for certain parts of the safety program. However, the authority to act must be delegated along with responsibility. While authority always starts with those in the highest administrative levels, it eventually must be passed down to middle and line management to achieve the desired results. If safety professionals, safety and health committees, department heads, supervisors, foremen, and employee representatives are to conduct a vigorous and thorough safety program, and if they are to accept and assert the authority delegated to them when circumstances warrant it, they must be fully confident that they have administrative support.

Management must understand that, although it can assert authority, it may encounter resistance unless it has enlisted employee support from the earliest stages of the program. If supervisors, employees, and their representatives are not aware of the reasons for and the benefits of an effective safety program, they may resist any changes in their methods of operation and instruction and may do as little as possible to assist the overall program effort.

Management must insist that safety and health information is an integral part of training, methods, materials, and operations. It must guarantee a system where loss control is considered an important part of equipment purchase and process design, operation, preventive maintenance, and layout and design. It must make sure that effective fire prevention and protection controls exist. Management also is responsible for informing subcontractor, at the time of negotiating the contract, of all applicable safety and health standards. Management must see that subcontractor comply fully with company and other safety regulations.

Management is required to safeguard employees' health be seeing to it that the work environment is adequately controlled. They must be aware that those operations producing airborne fumes, mists, smoke, vapors, gases, dust, noise, and vibration have the capacity to cause impaired health or discomfort among their workers. Management, must be aware that occupational illnesses beginning in the workplace can take their toll later, even after a worker retires. To protect the future of its employees, management must maintain an effective industrial hygiene monitoring system.

Management should also provide meaningful criteria to measure the success of the safety program and to provide information upon which to base future decisions. It must decide the safety program goals for reduced accidents, injuries, and illnesses, and their associated losses.

There are many concrete ways management can show evidence of its commitment to safety: managing safety and health programs, attending safety and health meetings, periodic walk-through inspections, reviewing and acting upon accident reports, investigating accidents, reviewing safety records through conferences with department heads and joint employee-management committees, providing awards, and by matting a good example. Management must also audit the program to be certain that it was not lost in the management chain. Department heads, supervisors, foremen, and employee representatives are in strategic positions within the organization to implement safety policies. Their leadership and influence should ensure that safety and health standards are enforced and upheld in each work area and that standards and enforcement are uniform throughout the workplace.

What are some responsibilities of department heads? They make certain that materials, equipment, and machines slated for distribution to their areas are hazard free or that adequate control measures have been provided. They make certain that equipment, tools, and machinery are being used as designed and are properly maintained. They keep abreast of accident and injury trends occurring in their areas and take proper corrective action to reverse these trends. They investigate all accidents occurring in their areas and take proper corrective action to reverse these trends. They investigate all accidents occurring within their jurisdiction. They should see to it that all safety and health rules, regulations, and procedures are enforced in their departments. They require that a hazard analysis be conducted for certain operations, particularly those that they regard as dangerous, either from past accident history or from their perception of accident potential. They require that hazard recognition end control information be included in instruction, training, end demonstration sessions for both supervisors and employees. They actively participate in and support the safety and health committee and follow up on its recommendations.

Supervisors, foremen, and employee representatives carry great influence. With their support, top management can be assured of an effective safety and health program, supervisors have a moral and professional responsibility to safeguard, educate, and train those who have been placed under their direction. Thus, they are generally responsible for creating a safe, healthy work setting and for integrating hazard recognition and control into all aspects of work activities. By their careful monitoring, they can prevent accidents.

For all practical purposes supervisors, foremen, and employee representatives are the eyes and ears of the workplace control system. On a day-to-day basis, they must be aware of what is happening in their respective areas, who is doing it, how various tasks are being performed, and under what conditions. As they monitor their areas, they must prevent accidents from occurring. If, despite their controls, they see danger, they must be prepared to intervene in the operation and take immediate corrective action. What are the chief safety and health responsibilities of supervisors?

They train and educate workers in safe working methods and techniques. They should be certain that employees understand the properties and hazards of the materials they store, handle, or use. They make sure that employees observe necessary precautions, including proper guards and safe work practices. They furnish employees with the proper personal protective equipment, instruct then in its use, and make certain it is worn.

Supervisors demonstrate an active interest in and comply with loss control policy and safety and health regulations. They actively participate in and support the safety and health committees. They supervise and evaluate worker performance, with consideration given to safe behavior and work methods. They should first try to convince employees of the need for safe performance, and then, unpleasant though it may be, they should administer appropriate corrective action when health and safety rules are violated. Correcting employees requires tact and good judgement. Enforcement should be viewed as education rather than discipline. However, if a supervisor feels that a worker is deliberately disobeying rules or endangering his or her life and the lives of others, then prompt and firm action is called for. Laxity in the enforcement of safety rules undercuts the entire safety and health program and allows accidents to happen. Supervisors monitor their area on a daily basis for human, situational, and environmental factors capable of causing accidents. They should make sure that meticulous housekeeping practices are developed and used at all times. They should correct hazards detected in their monitoring or report such hazards to the persons who can take corrective action. They should investigate all accidents occurring within their areas to determine causes.

Foremen and employee representatives share much of the responsibility for safety and health with upper management. Their job is to inspect, detect, and correct. What are the specific responsibilities of foremen and representatives? They should encourage other workers to comply with the organization's safety and health regulations. They detect safety violations and hazardous machinery, tools, equipment, and other implements. They take corrective action when possible and report the hazard to the supervisor, along with the corrective action taken or still required. They may participate in accident investigations. They represent the interests of the workers on the safety and health committee. Practical training aids for employee representatives, foremen, and supervisors do exist. The Council's "Supervisors' Safety Manual" and accompanying "Supervisors' Development Program" have been widely accepted by the industry. The Council's Joint Safety and Health Committee Training Course is also a good training tool and review for safety committees.

Housekeeping and preventive maintenance. This can be regarded as two sides of the same coin.

The safety program can succeed if housekeeping and maintenance are not seen as important to the effort.

Good housekeeping reduces accidents, improves morale, and increases efficiency and effectiveness. Most people appreciate a clean and orderly workplace where they can accomplish their tasks without interference and interruption. An industrial organization, by its very nature, contains tools that must be kept clean. In operations, workers use flammable substances and materials requiring special storage and removal. The processes generate dust, scrap metal filings and chips, waste liquids, scrap lumber, and countless other by-products that must be disposed of.

Housekeeping is a continuous process involving both workers and custodial personnel. A good housekeeping program incorporates the housekeeping function into all processes, operations, and tasks performed in the workplace. The ultimate goal is for each worker to see housekeeping as part of their job performance and not as an extra task that someone else should do. When the workplace is clean and orderly and housekeeping becomes a standard part of operations, less time and effort will be spent keeping it clean, making repairs, and replacing equipment, fixtures, and the like. When the worker can concentrate on required tasks without excess scrap material, tools, and equipment interfering with work, operations will be more efficient and the product of higher quality. Time will be used for work, not searching for tools, materials, or parts. When a plant is clean and orderly as well as safe, employee morale is heightened.

When everything has an assigned place, there is less chance that materials and tools will be taken from the plant or misplaced. In a few moments before quitting time, the supervisor can determine what is missing. Different colors of paint can be applied to tools to identify the department to which they belong. Tool racks or holders should be painted a contrasting color as a reminder to workers to return the tools to their proper places. When stored on a rack, the space directly behind each tool should be painted or outlined in color to call attention to a missing tool. Savings and efficiency are increased when workers treat materials with the care they deserve by minimizing spillage and scrap, saving pieces of material for use in future projects, and returning even snail parts to their storage area. When aisle and floor space is uncluttered, covenant within the plant is safer, and workers can easily clean and maintain their machinery and equipment. When the plant has adequate work space and when oil, grease water, and dust are removed from floors and machinery, workers are less likely to slip, trip, or fall, or inadvertently cone into contact with dangerous parts of machinery.

Chances of fires also are minimized when a workplace is kept free from accumulations of combustible materials that can burn upon ignition or, in the case of certain material relationships, spontaneously ignite. Furthermore, an orderly plant permits easy exit by keeping exits and aisles leading to exits free from obstructions. A neat, well organized workplace also makes it easier to locate and obtain fire emergency and extinguishing equipment.

Preventive maintenance is the orderly, uniform, continuous, and scheduled action to prevent breakdown and prolong the useful life of equipment and buildings. Preventive maintenance is a shared responsibility. Workers caring for tools and equipment accomplish specific maintenance tasks. Other maintenance duties, such as oiling, tightening guards, adjusting tool rests, and replacing wheels, are routinely performed by workers. Advantages to be gained from preventive maintenance includes safer working conditions, decreased downtime of equipment because of breakdown, and increased life of the equipment.

Satisfactory production depends on having buildings, equipment, machinery, portable tools, safety devices, and the like in operating condition and maintained so that production activities will not be interrupted while repairs are being made or equipment replaced. Preventive maintenance prolongs the life of the equipment by ensuring its proper use. When tools are kept dressed or sharpened and in satisfactory condition, the right tool will be used for the job. When safe and properly maintained tools are issued, workers have an added incentive to give the tools better care. When repairs are made quickly, workers do not need to improvise — for example, using a strip of metal as a screwdriver or crowbar. Sound, efficient maintenance management anticipates Machine and equipment deterioration and sets up overhaul procedures designed to correct defects as soon as they develop. Such a repair and overhaul system obviously requires close integration of maintenance with inspection.

Preventive maintenance has four main components:

1. Scheduling and performing maintenance functions.
2. Keeping records of service and repairs.
3. Repairing and replacing equipment and equipment parts.
4. Providing spare parts control.

Maintenance schedules can be set up on several different bases. Factors usually considered include:

1. Manufacturer's recommendations.
2. Age of the machine.
3. Number of hours per day the machine is used.
4. Past experience.
5. Machine changes with use.

The manufacturer's specifications contain standards that need to be maintained for safe and economical use of the Machine. These specifications give maintenance personnel definite guidelines to follow. Examples of scheduled activities include lubricating each piece of equipment; replacing belts, pulleys, fans, and other parts; and checking and adjusting brakes.

Management should keep two types of maintenance records: The first is a maintenance service schedule for each piece of equipment. This schedule indicates the date the equipment was purchased or placed in operation, its cost (if known), where it is used, each part to be serviced, the kind of service required, the frequency of service, and the person assigned to the servicing. Each piece of equipment also requires a repair record, which includes an itemized list of parts replaced or repaired and the name of person who did the work.

In addition to scheduled adjustments and replacements, maintenance personnel must repair malfunctioning or broken equipment in accordance with the manufacturer's specifications. Sometimes equipment must be sent back to the manufacturer or its service representative for repair.

Maintenance personnel should be aware of their limitations and recognize that their experience and expertise are not sufficient to do all repairs. Those assigned repair responsibilities require special safety training. Many of the jobs to be performed include testing or working on equipment with guards and safety devices removed. Therefore, a statement of necessary precautions should accompany the repair directive. Maintenance personnel along with others have a responsibility to lock out and tag defective equipment, or simply tag if the equipment, such as a ladder, cannot be locked out.

The second maintenance record itemizes spare parts stock. Management should conduct routine surveys of spare parts requirements. To keep needed repair parts on hand, management should schedule review and reordering or stocked spare parts. When maintenance personnel keep purchasing agents informed of their anticipated stock needs they help to reduce or eliminate downtime that occurs while workers wait for parts.

The difference between a mediocre maintenance program and a superior one is that while the former is aimed at maintaining facilities, the latter is designed to improve them. If conditions are good, a mediocre program will keep them that way but will not make them better; if conditions are not good, a mediocre program will not improve them. Preventive maintenance, on the other hand, is a program of mutual, support that creates safe conditions, eliminates costly delays and breakdowns, and prolongs equipment life.

Employees make the safety and health program succeed. Well- trained and educated employees are the greatest deterrent to damage, injuries and health problems in the plant or establishment.

What are the specific ways a safety program can be rooted in employee involvement and concern? Employees can observe safety and health rules and regulations and work according to standard procedures and practices. They can recognize and report to the foreman or supervisor hazardous conditions or unsafe work practices in the plant. They can develop and practice good habits of hygiene and housekeeping. They can use protective and safety equipment, tools, and machinery properly. They can report all injuries or hazardous exposure as soon as possible. Employees can help develop safe work procedures and make suggestions for improving work procedures. Management should encourage employees to participate on safety committees.

What shapes an employee's attitude toward safety? From the first day an employee goes to work, the employee starts to form attitudes about the job and organization. Substantial if subtle influence is exerted by the attitudes they observe in management, supervisors, and fellow workers. If these individuals regard safety as vital to an effective operation, a mark of skill and good sense, if they all participate actively and cooperatively in the safety program, if employees are recognized for having good safety records, then the employee will regard safety as something important, not as window dressing or an empty policy to which others pay lip service.

Introduction to the safety program should come on the employee's first day on the job. A three-pronged approach is suggested:

1. Coverage of general company policy and rules, discussion of various benefit programs, for example, hospitalization, pension plan, holidays, sick leave. The personnel department,

usually is responsible for giving this information immediately after the new employee is added to the payroll.

2. Discussion of general safety rules and the safety program. This part of the program should be the responsibility of a safety professional. The company's safety handbook should be given to the new employee, and the company's safety policy statement should be explained. The reasons behind the general safety rules should be explored with the employee, who is more likely to follow rules when they are understood.
3. Explanation of specific safety rules that apply to the new employee's department. At this point the supervisor's role overlaps with that of the safety professional. The supervisor can show how some hazards have been eliminated while others, which could not be designed out of the operation, are guarded against. This discussion provides an opportunity to talk about safe work practices and emergency procedures as well as to show how engineering controls and personal protective equipment can further reduce the effects of the hazard, the employee's responsibilities for safety and health also must be stressed. These responsibilities include reporting all accidents and incidents, checking equipment and tools before use, operating equipment only with proper authorization and prior instruction, and asking questioning when the procedure is unclear.

Those responsible for purchasing items for organizations are to reduce hazards associated with operations. The purchasing department has much latitude in selecting machinery, tools, equipment, and materials used in the organization. In maintaining standards of quality, efficiency, and price, the purchasing department must make certain that safety has received adequate attention in the design, manufacture, and shipping of items.

Depending upon the company organization, other departments such as safety, engineering, quality control, maintenance, industrial hygiene, and medical should indicate to the purchasing department what equipment and materials meet with their approval. The purchasing department is responsible for soliciting such guidance and direction.

First, the purchasing agent must make certain all items comply with regulatory standards. A statement to this effect must be part of the purchase order. Purchasing agents also will be guided by the standards of ANSI, Canadian Standards Association (CSA), and other standards and specifications groups; products approved or listed by such agencies as UL and the NFPA; and recommendations by such agencies as the NSC, insurance carriers or associations, the Factory Mutual System in its Factory Mutual Handbook of Industrial Loss Prevention and Loss Prevention Data," and trade or industrial organizations.

Second, the purchasing agent must make certain tools, equipment, materials, machinery, and chemicals are purchased with adequate regard for safety. This requirement applies even to ordinary items such as boxes, cleaning rags, paint, and common hand tools. It is essential to compare safety features among the various brands when purchasing personal protective equipment and larger items, especially machines. Sometimes the cost of an adequately guarded Machine seems out of proportion to that of an unguarded Machine to which makeshift guards can be added. But experience has repeatedly proven that the best time to eliminate or minimize a hazard is in the design stage. Safeguards that are integral parts of a Machine are the most efficient and durable.

Third, the purchasing agent must be cost conscious, realizing that every accident has both direct and indirect costs. The agent should understand that the organization cannot afford bargains that later result in accident losses and occupational disease.

The roles of various professionals in the loss control program are described in National Safety Councils "Fundamentals of Industrial Hygiene," 3rd ed. The following discussion, therefore, describes briefly the responsibilities of the safety and health professional, the industrial hygienist, and staff medical personnel.

Safety and health professionals (loss control specialists). To ensure the effectiveness of the safety program, management usually places program administration in the hands of a safety director or manager of safety and health. The number of full-time safety professionals is increasing as the nature of their duties is better understood.

To administer a safety program effectively requires considerable training and many years of experience. A safety program has many facts: occupational health, product safety, Machine design, plant layout, security, damage control, fire prevention. Safety as a profession combines engineering, management, preventive medicine, industrial hygiene, and organizational psychology. It requires knowledge of system safety analysis, job safety analysis, job instruction training, human factors engineering, biomechanics, and product safety. The professional must have thorough knowledge of the organization's equipment, facilities, manufacturing process, and workers' compensation, and must be able to communicate and work with all types of people. Other desirable traits are the ability to see both management and employee viewpoints, and the skills of a good trainer.

The passage of the 1970 OSH Act requires that certain safety standards be met and maintained. Generally, organizations with moderate or high hazards and/or employing 500 or more persons need a full-time safety professional. However, the nature of the operation may indicate the need for a full-time professional, regardless of the number of people employed.

The safety professional whether called a safety engineer, safety director, hazard control specialist, loss control manager, a safety and health professional, or some other title, normally functions as a specialist on the management level. The loss control program should enjoy the same position as other established activities of the organization, such as sales, production, engineering, or research. Its budget reflects top management's commitment to the safety and health of its employees and includes the safety professional's salary, the salary of staff to help him or her, travel allowance, cost of safety equipment, cost of training and continuing education, and other related items. Safety professionals must define the needs of their program and, according to priorities, make short-range and long-range budget projections. With such projections in hand, they can present their needs to those with fiscal responsibility and stand a better chance of acquiring the resources to make their programs work.

In general, the safety professional advises and guides management, supervisors, foremen, employees, and such departments as purchasing, engineering, and personal on all matters pertaining to safety. Formulating, administering, monitoring, evaluating, and improving the accident prevention program are other responsibilities of the safety and health professional.

The safety professional usually investigates serious accidents personally or through the staff, reviews supervisors' accident reports, checks corrective actions taken to eliminate accident causes, makes necessary reports to management, and maintains the accident report system, including files. The safety and health professional recommends safety provisions in:

1. Plans and specifications for new building construction.
2. Repair or remodeling of existing structures.
3. Safety equipment.
4. Designs of new equipment.
5. Processes, operations, and materials.

The safety professional provides (or cooperates with the training supervisor to provide) safety training and education for employees. The safety staff also take continuing education courses and attend professional meetings like the National Safety Congress and Exposition.

Further, the safety professional makes certain that line managers know and understand their responsibility to comply with regulatory standards relating to safety and health and that standards, whether mandatory or recommended, are met; accepts responsibility for preparing the necessary reports (for example, OSHA Forms 100, 102, and 200) for management.

On jobs involving subcontractors, the safety professional informs line management of their responsibility to inform each subcontractor of relevant hazard control responsibilities; this is often spelled out in the contract. The safety professional also supervises disaster control, fire prevention, and firefighting activities when they are not responsibilities of other departments and stimulates and maintains employee interest in and commitment to safety.

Progressive management will give the safety professional the authority to order immediate changes on fast-moving and rapidly changing operations or in cases where delayed action could endanger lives (as in construction, demolition, or emergency work, fumigation, and some phases of manufacturing explosives, chemicals, or dangerous substances). Where the safety professional exercises this authority, it is done with discretion by accepting accountability to management for errors in judgment. However, the professional realizes that erring on the side of caution is more easily justified than taking unnecessary risks with worker health and safety.

Industrial hygienists. The industrial hygienist is trained to anticipate, recognize, evaluate, and control health hazards particularly chemical, physical, biological, and ergonomic agents that exist in the workplace and have injurious effects on workers. Specialists work in fields such as toxicology, epidemiology, chemistry, ergonomics, acmustics, ventilation engineering.

What are the specific responsibilities of the industrial hygienist in the hazard control program? The industrial hygienist recognizes and identifies those chemical, physical, and biological agents that can adversely affect the physical and mental health and well-being of the worker; measures and documents levels of environmental exposure to specific hazardous agents; evaluates the significance of exposures and their relationship to occupationally and environmentally induced diseases; establishes appropriate controls and monitors their effectiveness; and recommends to management how to correct unhealthy (or potentially unhealthy) conditions. The roles of the industrial hygienists are detailed further in the Councils "Fundamentals of Industrial Hygiene."

Occupational health services vary greatly from one organization to another, depending on number of employees, nature of the operation, and commitment of the employer. One location may include a full-tine physician, nurses, and technicians, with treatment rooms and a dispensary. Another may have only the required first aid kit and a person trained to render first aid and Cardiopulmonary Resuscitation (CPR). The snail operation has special need for persons thoroughly trained in first aid and CPR to take care of the employees on all shifts. Sometimes small facilities in the same locality share the services of a qualified physician on either a part- or full-time basis.

How much responsibility a part- or full-time nurse assumes generally depends on the availability of a licensed physician. It is not unusual for an occupational nurse to be solely and completely responsible for the occupational safety and health program with medical direction from a physician who rarely visits the workplace.

The American Association of Occupational Health Nurses in 1977 defined occupational health nursing as "the application of nursing principles in conserving the health of workers in all occupations. It involves prevention, recognition, and treatment of illness and injury, and requires special skills and knowledge in the fields of health, education, and counseling, environmental health, rehabilitation, and human relations." In addition to being responsible for health care, the occupational nurse may on occasion monitor the workplace, perform industrial hygiene sampling, net as a consultant on sanitary standards, and be responsible for health education.

Occupational Health Programs, deals specifically with the duties of the industrial physician, nurse, and first aid attendants. Regardless of size, health programs share common goals: to maintain the health of the workforce, to prevent or control diseases and accidents, and to prevent or at least reduce disability and resulting lost time.

Staff medical personnel are a vital part of the total health and safety program. Their most obvious contribution is to provide emergency medical care for employees who are injured or become ill on the job. An outgrowth of this responsibility is to provide follow-up treatment of employees suffering from occupational disease or injuries. For many, periodic examinations are required by OSHA regulations. A further outgrowth is medical personnel's promotion of health education programs for employees and their families some ways that medical personnel can encourage the health and well-being of employees include assisting with problems of alcohol and drug abuse; keeping immunizations up to date; and promoting mental health, weight control, and regular exercise.

Medical personnel foster a healthful environment by occasional tours of the workplace. Such tours also familiarize medical personnel with the materials and processes used and procedures performed by the employees under their care.

Another way that medical personnel are important to the health and safety organization is in the placement of employees. Proper placement matches each worker tot the right job. Often a replacement examination gives the medical staff an opportunity to recommend where a prospective worker is to be assigned, This placement should take into consideration the physical and mental capacities of the employee so that no one is subjected to unnecessary safety and health risks.

HANDBOOK OF FIRE PROTECTION ENGINEERING

K. EHTESHAMI

FEEDBACK FORM

1. Which topics from your examination point of view are inadequately or not covered in this book?

____

____

2. Is there any misprints/ mistakes/ factual inaccuracy in this book? If so, please specify.

____

____

3. What is your assessment of this book as regards the presentation of the subject-matter, expression, precision and price in relation to the other recommended books available on this subject?

____

____

4. Which competing books you regard as better than this? Please mention their author and publishers.

____

____

5. Any further suggestions/comments you would like to make for the improvement of the book?

____

____

PS: If needed, please attach a separate sheet.

Name : ____________________

College : ____________________

Address : ____________________

Email : ____________________

Mobile : ____________________

Please mail the feedback form at:

KHANNA BOOK PUBLISHING CO. (P) LTD.

4C/4344, Ansari Road, Daryaganj, New Delhi-110002

You can also email this feedback form at contact@khannabooks.com or Whatsapp at +91 – 99109 09320

BEST FEEDBACK WILL BE REWARDED SUITABLY